INTERNATIONAL CENTRE FOR MECHANICAL SCIENCES

COURSES AND LECTURES - No. 225

RANDOM EXCITATION OF STRUCTURES BY EARTHQUAKES AND ATMOSPHERIC TURBULENCE

EPITED BY
H. PARKUS
TECHNICAL UNIVERSITY OF VIENNA

SPRINGER-VERLAG WIEN GMBH

Originally published by Springer-Verlag Wien-New York in 1977

ISBN 978-3-211-81444-4 ISBN 978-3-7091-2744-5 (eBook)
DOI 10.1007/978-3-7091-2744-5

P r e f a c e

Earthquakes are not predictable in a deterministic sense. It is this fact that contributes largely to their disastrous effects. The same, although to a lesser degree, holds true for wind effects. A meaningful analysis of these phenomena must, therefore, take their random nature into account.

Probabilistic methods in the theory of structures are not a regular part of engineering education. It was, therefore, decided by the scientific council of CISM to organize a series of courses on the subject in Udine during the third week of July 1976.

While preparations were still underway, a severe earthquake shook the region of Friuli in the immediate vicinity of Udine. More than 1000 people lost their lives and heavy damage was incurred. This extremely unfortunate event, however, stimulated interest in the CISM courses. The number of participants reached an unexpectedly high level.

The present volume contains the courses and lectures presented at Udine. I would like to thank the authors for their efforts in presenting the lectures and preparing the manuscripts for publication. My thanks are also due to my dear friend Prof.W. Olszak, rector of CISM, for his continued advice and encouragement.

I am particularly grateful to Dr.R. Grossmayer for taking care of most of the editorial work, and to my secretary, Ms.A. Kienast, for the careful typing of the manuscript and the expert rendering of the equations.

H. Parkus

C O N T E N T S

AN APPROACH TO CHARACTERIZING, MODELING AND ANALYZING EARTHQUAKE EXCITATION RECORDS

by F. KOZIN

ASEISMIC RELIABILITY AND FIRST-PASSAGE FAILURE

by R. GROSSMAYER

APPLICATIONS OF DIGITAL SIMULATION

OF GAUSSIAN RANDOM PROCESSES

by M. SHINOZUKA

STRUCTURAL RESPONSE UNDER TURBULENT FLOW EXCITATIONS

by Y.K. LIN

SEISMIC SAFETY ASSESSMENT

ERIK H. VANMARCKE
Department of Civil Engineering
Massachusetts Institute of Technology

FOREWORD

The emphasis in these notes is on the use of random vibration analyses to predict the response of structures to earthquake ground motion. Random vibration analysis has as its aim the prediction of the probability distribution of a dynamic response parameter of interest in terms of the dynamic properties of the structure and a statistical description of the earthquake. The most convenient ground motion representation for this purpose is in terms of a spectral density function and an equivalent duration of strong-motion shaking. This representation is discussed in detail in Section 3 of these notes.

A blending of the theories of structural dynamics and probability has been underway for some time, but only recently have the major obstacles preventing its useful application to seismic analysis of general linear systems been cleared. One obstacle has been the lack of a workable methodology for dealing, in probabilistic terms, with the transient nature of the ground motion and the structural response, and another (obstacle) the difficulty in dealing with maximum values as principal response parameters.

Earthquake engineers are interested not in the details of the response motion but in a few response parameters which facilitate assessment of the system's performance during an earthquake. The performance of linear systems is most often evaluated in terms of the maximum response which is compared with the allowable (damage or failure) response value. In the broader context of overall seismis safety assessment of structures, random vibration seeks to provide the conditional system response given a (rough) description of the ground motion. To determine the "unconditional" seismic safety, random vibration results must be combined with the output of a seismic risk analysis and with information about the structures' dynamic properties and resistance. Methodology of seismic risk analysis is presented in Section 1 of these notes.

To put the random vibration approach to seismic response evaluation in proper perspective, a section of the notes reviews conventional procedures of seismic dynamic analysis. Section 2 considers both linear elastic and inelastic systems, and reviews procedures based on the response spectrum and on time integration of recorded or simulated earthquake accelerograms.

Random vibration analyses hold a clear advantage over other procedures in that they yield information about the <u>distribution</u> of structural response, allowing direct assessment of the probability of exceeding intolerable response levels. Moreover, the solutions often take a simple analytical form, which facilitates incorporating these results in an overall seismic safety analysis. Presently, the approach is practical for linear systems <u>only</u>, however. Some results exist for simple one-degree inelastic systems[1] and research is underway at M.I.T.[2] aimed at predicting the seismic response of inelastic shearbeam structures by approximate random vibration analysis.

In these notes, the types of systems treated probabilistically are <u>linear</u> one-degree and multi-degree-of-freedom systems. In each case, the solution is stated in terms of the response $y_{s;p}$ in which the subscripts refer to an exceedance probability p and the strong-motion duration s. The response "fractile" is expressed as follows

$$y_{s;p} = r_{s;p}\, \sigma_y(s)$$

where $\sigma_y(s)$ = standard deviation of the linear system response at time s

$r_{s;p}$ = peak factor which relates standard deviation to the response level not exceeded with probability p

Section 4 is devoted to determining $\sigma_y(s)$ for one-degree and multidegree linear systems. For one-degree systems, the stationary (corresponding to $s \to \infty$) response variance is derived first, and the required modifications to deal with the transient nature of both the excitation and the response are discussed subsequently. A general approach to estimating the peak factor $r_{s;p}$ for linear systems is described in Section 5. The approach is of necessity approximate, as no known exact solutions to the so-called "first-passage" problem presently exist. The lecture notes of Dr.Grossmayer in this volume shed further light on this problem.

Troughout the notes, the emphasis is on concepts and methodology, rather than on specific applications or numerical examples. Additional random vibration results dealing with uncoupled systems and with simple nonlinear structures, as well as some numerical examples of the results presented herein, may be found in a related publication by the writer[1].

1. ASSESSMENT OF THE EARTHQUAKE THREAT AT A SITE

Future earthquakes in the area surrounding the site of a structure are uncertain in their times of occurrence, their sizes and their epicentral locations. For the purpose of predicting what kind of ground motion it will generate at the site, an earthquake is often characterized by its Richter magnitude, M, and the distance from focus to site, R, or the epicentral distance, D. The meaning of these symbols is illustrated in Fig.1.1. Another measure of the size or strength of an earthquake is the Modified Mercalli Intensity at the epicenter or the epicentral intensity, I_o.

Each earthquake causes ground shaking at the site and generates some time history of motion at the point on the surface of the ground at the base of the structure. The most commonly used parameters of the motion-at-the-site are: the site intensity (Modified Mercalli scale), I; the peak ground acceleration, a; the peak ground velocity, v; the peak ground displacement, d, and the duration of the motion, s. These ground motion parameters can be related (rather imperfectly) to earthquake size and source-to-site distance through attenuation laws which are discussed subsequently. Also various empirical relations exist between site intensity and peak ground motion parameters and between the epicentral intensity I_o (in MM units; see Fig.1.1), and the magnitude, M (Richter scale). For details, the reader is referred to a textbook on Earthquake Engineering, such as Newmark and Rosenblueth[3].

The motion of the ground causes the structure resting on it to vibrate, thus generating time histories of structural response, e.g., displacements, stresses, or strains. For elastic structures, the response parameter which has most practical value is the maximum response. In particular, for a simple one-degree structure with natural frequency ω_n and damping ratio ζ, this quantity is usually represented by one of the following symbols which are all related to the relative displacement y(t):

Peak Response: S_d = peak relative displacement = $\max |y(t)|$

$$S_v = \omega_n S_d = \text{pseudo-velocity}$$

$$S_a = \omega_n^2 S_d = \text{pseudo-acceleration}$$

The pseudo-velocity is approximately equal to the maximum relative velocity of the structure, while the pseudo-acceleration is close to the maximum absolute acceleration of the structure. Response spectra are plots of the above given quantities as functions of natural frequency for different values of damping. These plots have been found to be very sensitive to the intensity (I) or to one of the peak amplitudes (a, v or d) of the ground motion. In fact, it has been recommended[4] that a point with coordinates (ω_n, ζ) on the response spectra can be determined simply by multiplying an appropriate ground motion parameter (I, a, v or d) by a factor which depends only on the structural characteristics, ω_n and ζ. Of course, as with attenuation laws, these relationships are not perfect. Actual peak responses are unpredictable even when all ground motion parameters listed above are given. The study of the statistical properties of structural response when a probabilistic description of the input motion is available is the object of random vibration theory. Its application to earthquake engineering problems is increasing in importance, and a major portion of these notes is devoted to this topic.

Our goal is to examine the structural response problem in a larger context, linked to the problem of predicting earthquake occurrence and to the problem of specifying the characteristics of ground motions to be used in analysis and design of a structure at a site. The sequence of "events" which take place when an earthquake happens are schematically depicted in Fig.1,2. Successive earthquakes affecting the structure at different times during its operational life will cause a sequence of peak structural responses. We hypothesize that the structure will perform satisfactorily only if all peak responses lie below its design response or resistance.

Modeling Earthquake Occurrence

On the basis of a statistical study of earthquake occurrence in time and space, Gutenberg and Richter[5] have found that an approximate linear relationship exists between log N(M) and the (Richter) magnitude M, where N(M) denotes the average number of earthquakes per year greater than or equal to M.

$$\log N(M) = A + BM \tag{1.1}$$

It follows that

$$N(M) = C\, e^{-BM} \tag{1.2}$$

The parameter C is equal to the product of the size and to the average seismicity of the ground area under consideration, and B is a constant which characterizes the rate of exponential decay in Eq.1.2. Values of B close to 2 are found in most parts of the world[6]. Eq.1.1 does not fit existing data for very large magnitudes. There appears to be an upper bound between 8.7 and 8.9 on the Richter scale.

For sufficiently large areas the constants in Eq.1.2 can be reasonably accurately estimated. The problem with taking large areas is that in the process of averaging occurrence rates, not enough attention is paid to the relative seismicity of active and inactive regions within the area. On the other hand, for a smaller area (as one often deals with in a seismic risk analysis at a site), the parameter estimation problem becomes very difficult*). Usually, only scattered historical information is available, and it needs to be integrated with whatever information geologists can supply about the location of active faults and pertinent subsurface conditions within the area.

Attenuation Laws

There is a strong correlation between what happens at the source and at the site, at the base of structure. Generally the intensity of the motion-at-the-site decreases with increasing distances between the site and the source. When distances are equal, earthquakes with larger

*) The parameter B in Eq.1.2 is often taken to be the same for all areas surrounding the site which are identified as potential sources of earthquakes. Only the parameter C (in Eq.1.2) needs to be estimated for each area.

magnitudes usually generate larger site intensities and ground motion amplitudes. This is reflected in so-called attenuation laws which are based primarily on instrumental and other observed data. For example, using actual strong-motion records from firm ground sites in the western United States, Esteva and Rosenblueth[7] determined that the peak ground acceleration, a, is related to the magnitude, M, and the modified focal distance, R, in the following way*)

$$a = 2000\ e^{0.8M}\ R^{-2} \tag{1.3}$$

Many other proposals have been made, all based on limited data which shows a large amount of scatter. Attenuation laws relating the site intensity, I, to the intensity in the epicentral region, I_o, and the epicentral distance, D, are also available, but will not be discussed here.

Seismic Risk Analysis

The method of seismic risk analysis developed by Cornell[8] combines information about times of occurrence of earthquakes, areal distribution of seismicity (complemented

---

*) The acceleration is in units of cm/sec^2. Esteva (1967) shows that in order to increase the accuracy of the law at close distances, R should be computed by using $\sqrt{D^2 + h^2 + 20^2}$, in which D is the epicentral distance and h the focal depth. All distances are measured in km.

by geological data wherever practical), and attenuation of motion intensity, to yield probabilistic statements about the seismic threat at a given site. The output of the analysis might be in the form of a peak acceleration-versus-mean return period plot as shown in Fig.1.3. A peak acceleration of 0.15g corresponds to a 200 year mean or average return period. This means that a ground motion with peak acceleration greater than or equal to 0.15g will occur on the average once every 200 years. Another interpretation is that the average annual number of earthquakes causing a peak ground acceleration $\geq$ 0.15g at the site equals 1/200 = 0.005. The latter number can further be interpreted as the probability that "a $\geq$ 0.15" will occur in any one year (all these interpretations are roughly equivalent for small snnual risks or large return periods).

The theory underlying seismic risk analysis is quite straightforward. First one attempts to identify the areas surrounding the site which are suspected of being potential sources of earthquakes. Each such source is then divided into smaller sub-sources (which can be regarded as point sources). For each sub-source, i, one determines:

a) the parameter C_i in the relationship (see Eq.1.2)

$$N_i(M) = C_i \; e^{-BM} = C_i \; e^{-2M} \tag{1.4}$$

in which $N_i(M)$ = the average annual number of

earthquakes with magnitude greater than or equal to M originating from sub-source i.

b) The focal distance R_i from the sub-source to the site.

At the site, interest focuses on the earthquakes which cause, say, peak ground accelerations greater than or equal to some fixed value $\underline{a}$. Let N_a denote the average annual number of such earthquakes. This number equals the sum of many contributions each due to a single one of the (non-overlapping) subsources. Therefore

$$N_a = \sum_{\text{all } i} N_{i,a} \tag{1.5}$$

where $N_{i,a}$ = the average annual number of earthquakes from sub-sources i which cause a peak acceleration $\geq$ a.

The magnitude M_a needed for a sub-source i quake to cause a peak acceleration $\underline{a}$ can be determined from the attenuation law, Eq.1.3

$$a = 2000\ e^{0.8M_a}\ R_i^{-2} \tag{1.6}$$

Solving Eq.1.6 for M_a one obtains

$$M_a = \log \left[R_i^2 \frac{a}{2000} \right]^{1/0.8} \tag{1.7}$$

Combining Eqs.1.4 and 1.7 yields

$$N_{i,a} = N_i(M_a) = c_i\ e^{-2M_a} = c_i R_i^{-5} \left[\frac{a}{2000} \right]^{-2.5} \tag{1.8}$$

The final step is to introduce Eq.1.8 into Eq.1.5

$$N_a = \left[\sum_{all\ i} C_i R_i^{-5}\right]\left[\frac{a}{2000}\right]^{-2.5} = G\left[\frac{a}{2000}\right]^{-2.5} \qquad (1.9)$$

in which G is a factor which depends on the geometry and relative activity of the various sources. Note that the relationship between N_a and a given by Eq.1.9 will be represented by a straight line on log-log paper. The mean return period T_a is simply the reciprocal of N_a.

$$T_a = \frac{1}{G}\left[\frac{a}{2000}\right]^{2.5} \qquad (1.10)$$

Recall that the acceleration is given in units of cm/sec^2, and T_a is in years. The result for a southern California site[9] is shown in Fig.1.3.

The results of a number of site risk analyses can be combined to construct Seismic Probability Maps or Seismic Regionalization Maps for a given area. Such maps can be valuable tools in siting and seismic design.

Some Other Factors Affecting Seismic Risk

The methods presented lead to a description of the seismic threat at a site in terms of the relationship between peak ground acceleration (or intensity, etc.) and the average return period. Randomness of occurrence times, sizes and locations of future earthquakes are accounted for, and widely used assumptions for the relative frequencies of magnitudes (Eq.1.2) and attenuation laws (Eq.1.3) are adopted. Some factors which can be quite important and have been studied, but which have not been dealt with in the above analysis, are the following:

a) There is significant scatter of observed ground motion data about the values predicted by the attenuation laws (which were fitted to data from many different sites and sources). For a summary of recent work, the reader is referred to Esteva[10], and Cornell and Vanmarcke[11].

b) Eq.1.2 grossly overestimates the number of earthquakes with very large magnitudes. One solution is to use Eq.1.2 only for magnitudes less than some specified upper bound, another is to use a quadratic rather than an exponential relationship between N(M) and M[12].

c) Seismic risk analysis results are quite sensitive to the parameter values chosen in Eqs.1.2 and 1.3. Since adequate information upon which to base the parameter estimation is frequently lacking, careful sensitivity studies become very important. For some examples and for an appraisal of the potential and the limitations of seismic risk analysis, the work of Donovan[13] is recommended.

2. CONVENTIONAL SEISMIC ANALYSIS PROCEDURES

The simplest seismic analysis procedures are based directly on response spectra. In analyzing linear elastic multi-degree systems by the response spectrum approach, the response spectrum is used to predict the peak response for each mode of the structure. The individual modal maxima are then combined, usually by the square root of the sum of the squares, to provide the expected peak response of the complete structure. This method gives no information about the degree to which actual responses might deviate from the predicted value. Similar approximate procedures have been proposed to predict, for example, the response of light equipment in buildings and of simple nonlinear systems directly from a set of specified response spectra.

Another common procedure involves step-by-step time integration based on one or more recorded or simulated accelerograms. Recorded ground motions which are reasonably representative of the type of motion to be expected at a site are often not available. Moreover, engineers are aware that, due to the stochastic nature of earthquake ground motion, the information obtained from a structural response analysis using a single record is quite unreliable. It is possible to generate by computer a set of artificial motions which cannot be distinguished, as regards over-all statistical properties, from actual recorded ground motions. Frequently, these simulated motions are "designed" to have computed response spectra which oscillate around the specified expected response spectra for the site. By calculating structural responses for each record in a sample, one can construct relative frequency curves for a response parameter of interest, or at least obtain reasonably stable estimates of its average and perhaps of its standard deviation. The step-by-step integration procedure has the advantage of being generally applicable, e.g., in the case of complex nonlinear systems which other procedures cannot handle, but it is often expensive and timeconsuming.

Throughout these notes, attention is restricted to investigating the dynamic response under only one component of ground motion. For linear systems, the analysis can relatively easily be extended if the earthquake excitation is represented by three statistically independent translational components of ground motion. Recent evidence[14] supports the use of such a model.

Linear Multi-Degree Systems

The first step in the seismic analysis of a structure

or a soil-structure system is to construct a dynamic model, frequently a lumped-parameter model whose parameters are the elements of the mass and stiffness matrices. These may be determined by any one of several conventional procedures (see, for example[15]). The natural frequencies and shapes of the normal modes can then be determined by solving the eigenvalue problem. In the normal mode method, the n-degree system response at a predetermined point B on the structure is expressed in terms of the modal coefficients c_k and the generalized modal coordinates $y_k(t)$, $k = 1,2,\ldots,n$. In particular, the displacement at point B relative to the ground is:

$$y(t) = \sum_{k=1}^{n} \phi_{kB} \Gamma_k \, y_k(t) = \sum_k c_k \, y_k(t) \qquad (2.1)$$

Each component, $y_k(t)$, is the response of a one-degree system characterized by the (undamped) natural frequency ω_k and an assigned percentage of critical damping ζ_k. Also, $c_k = \phi_{kB} \Gamma_k$, where ϕ_{kB} = the characteristic shape ordinate for mode k at point B, and Γ_k = the participation factor in the kth mode. For more details about the normal mode technique, the reader is referred to a textbook on structural dynamics[15,16,17].

In complex systems where damping varies substantially in nature (hysteretic vs. viscous) or magnitude throughout the system, classical modal analysis is not strictly applicable. The difficulty with normal mode approximations lies in the assignment of modal damping[18]. For example, in soil-structure interaction analysis, the damping in part of the system is viscous while in other parts it is more nearly hysteretic. Roesset et al[19] and Bielak[20] have suggested methods for determining weighted modal damping. An alternative to the modal analysis approach is a solution by frequency domain analysis (e.g.,Sarrazin et al.[21].

The response spectrum approach is the simplest and most common way to estimate the maximum system response, say, the maximum relative displacement, due to a ground motion. The response spectrum is utilized to produce the maximum modal displacements S_{dk}. These are read from the response spectrum at the period ω_k and for the damping coefficient ζ_k. The maximum displacement at B (relative to the ground) is often predicted by the "root-sum-square" rule[22]:

$$S = \sqrt{\sum_{k=1}^{n} c_k^2 \; S_{dk}^2} \qquad (2.2)$$

Similarly, the modal maximum acceleration at mass point B on the structure are $A_{kB} = \left|\omega_k^2 \; S_{dk} \; c_k\right|$. When multiplied by the mass (at B), these accelerations are the equivalent static forces which may be used to compute the maximum stresses in the structure due to mode k. The modal stresses at a point may then be combined by the root-sum-square procedure to predict total maximum stress. Other rules have been suggested[23] which attempt to account for the interaction between modal contributions (which becomes more pronounced when modal frequencies are close or when the damping is high).

An alternative analysis procedure is based on the numerical solution of the modal equations of motion (by time- or frequency-domain methods). The output may be in the form of individual modal response functions $y_k(t)$ which can be superimposed using Eq.2.1. The maximum value of y(t) is then sought. This procedure can be repeated for several actual earthquakes or simulated ground motions to determine the range of possible maximum responses.

Inelastic Systems

In earthquake engineering perhaps more than in any other area of applied dynamics, it is necessary to face up to important nonlinear effects in the behavior of structures and soils. The stiffness and strength values appropriate at low amplitudes of motion and the mechanism of energy dissipation change as a function of the motion level. They also depend on past seismic action and other environmental conditions. For example, in reinforced concrete structures, cracking of concrete and slip and yielding of reinforcement tend to reduce significantly both stiffness and strength.

Also, natural frequencies and damping factors of real buildings are affected by difficult-to-estimate nonlinear effects attributable to nonstructural components and foundation soil.

Many types of nonlinearities can be dealt with satisfactorily using "equivalent linear" analyses. Linear dynamic properties are chosen to be compatible with expected response levels. This method is frequently used in studies of soil amplification and liquefaction[24]. Shear modulus and damping values are adjusted until they are compatible with computed strains in the soil.

A survey of a variety of nonlinear stiffness and damping effects is given by Crandall[25]. It appears that most types of relatively <u>mild nonlinearities</u> can be dealt with successfully by equivalent linearization techniques[26]. But these techniques do not succeed in capturing the essence of <u>strongly nonlinear</u> hysteretic behavior, i.e., energy dissipation through the development of drift and permanent set.

Many structural systems are designed to undergo plastic deformation during very severe but infrequent earthquake shaking. Design level seismic analyses based on <u>linear</u> dynamic models of buildings often predict elastic responses in excess of elastic capacities[27]. Building componenets which are ductile have far more resistance than the elastic analysis would indicate. What they lack in elastic strength is made up by their capacity to absorb energy through inelastic action. Limits on the values of member distortions are set by available ductility or excessive deflection or overall instability.

The most common model of nonlinear hysteretic stress-strain behavior is the elasto-plastic system shown in Fig.2.1. Newmark and Hall[4,28] have developed a simplified procedure for determining the relationship between the maximum response of an elasto-plastic system and the maximum response of the associated elastic system (see Fig.2.1b). The result is the <u>inelastic response spectrum</u> which consists of an acceleration spectrum and a (relative) displacement spectrum. Both can be plotted as a function of the (initial undamped) natural period for a specified level of the ductility ratio μ and the damping factor ζ. The ductility ratio μ is the ratio of the maximum deflection to the limit elastic deflection ($\mu = d_{max}/a$, see Fig.2.1).

Starting from the pseudo-velocity elastic response spectrum plotted on log-log paper, the inelastic response spectrum is developed as follows. In the period range where displacement or velocity are amplified, the inelastic displacement spectrum is identical to the elastic spectrum. The inelastic acceleration spectrum in these regions is obtained by dividing the elastic spectrum by the ductility ratio μ. In the amplified acceleration period range, the inelastic acceleration spectrum is obtained by dividing the elastic spectrum by $\sqrt{2\mu - 1}$, while at very high frequencies it is identical to the elastic response spectrum. The spectrum is completed by drawing a straight line from the constant acceleration line to the amplified acceleration line. Note that the inelastic displacement and acceleration spectra always differ by the factor μ. Further details and background material can be found in Newmark and Rosenblueth[3]. For example, starting from the elastic response spectrum represented by the dotted line in Fig.2.2, the inelastic response spectrum is derived for a damping of 2 percent and for a ductility ratio of 4. Response spectra for other nonlinear hysteretic systems have been obtained by Veletsos[29].

Many researchers[26,29,30,31] have obtained response statistics of simple nonlinear hysteretic systems through step-by-step integration procedures using recorded and simulated ground motions. Of course, these procedures can be extended to much more complex models of the structure's behavior in the inelastic range. The choice of an appropriate analytical model poses a major problem, though. Experimental work has revealed that component behavior is quite complex, including deterioration of stiffness and strength with number of cycles and in function of the level of deformation. A true-to-life inelastic analysis of a building frame, if possible, is clearly impractical. The most common model is a "shear building" with elasto-plastic or bilinear springs in each story. The ultimate capacity of each spring may be the sum of the ultimate column end moments divided by the story height.

Time-integration analyses based on inelastic models of real buildings[32] have revealed that the overall deformation of the building is close to that predicted based on an elastic model with the "initial" properties of the inelastic model, but that the maximum strains in the individual components may be larger by a factor of 6. Biggs and Grace[33] found that a crude estimate of the average inelastic interstory displacement can be made using elastic, first mode, interstory displacements. Dividing the latter by the average elastic limit interstory displacement gives an estimate of the average ductility ratio of the building. The pattern of inelastic action, i.e., the distribution of ductility versus height, shows large variations from earthquake to earthquake.

3. SPECTRAL REPRESENTATION OF EARTHQUAKE GROUND MOTIONS

A typical accelerogram has the appearance of a

transient, stochastic function[34,35] and it is not difficult to generate by computer many accelerograms which cannot be distinguished, as regards over-all statistical properties, from recorded ground motions. One commonly used method of numerical simulation of earthquakes is based on the fact that any periodic function can be expanded into a series of sinusoidal waves:

$$x(t) = \sum_{i=1}^{n} A_i \sin(\omega_i t+\phi_i) \tag{3.1}$$

A_i is the amplitude and ϕ_i is the phase angle of the ith contributing sinusoid. By fixing an array of amplitudes and then generating different arrays of phase angles, different motions which are similar in general appearance (i.e., in frequency content) but different in the "details", can be generated. The computer uses a "random number generator" subroutine to produce strings of phase angles with a uniform distribution in the range between 0 and 2π.

The <u>total power</u> of the steady state motion, x(t), is $\sum_{i=1}^{n} (A_i^2/2)$. Assume now that the frequencies ω_i in Eq.3.1 are chosen to lie at equal intervals $\Delta\omega$. Fig.3.1 shows a function $G(\omega)$ whose value at ω_i is equal to $(A_i^2/2\Delta\omega)$ so that $G(\omega_i)\Delta\omega = A_i^2/2$. Allowing the number of sinusoids in the motion to become very large, the total power will become equal to the area under the continuous curve $G(\omega)$, which is in effect the <u>spectral density function</u>. Formal definitions of $G(\omega)$ can be found in many textbooks[36,37]. $G(\omega)$ expresses the relative importance (i.e., the relative contribution to the total power) of sinusoids with frequencies within some specified band of frequencies. When $G(\omega)$ is narrowly centered around a single frequency, then Eq.3.1 will generate nearly sinusoidal functions as shown in Fig.3.2(a). On the other hand, if the spectral density function is nearly constant over a wide band of frequencies, components with widely different frequencies will compete to contribute equally to the motion intensity, and the resulting motions will resemble portions of earthquake records, as illustrated in Fig.3.2.(b). Of course, the total power and the relative frequency content of the motions produced by using Eq.3.1 do not vary with time. To simulate in part the transient character of real earthquakes, the stationary motions, $x(t)$, are usually multiplied by a deterministic intensity function such as the "boxcar", trapezoidal[38], or exponential[39] functions shown in Fig.3.3.

Bycroft[40] and Brady[41] simulated "white noise" (for which $G(\omega)$ is theoretically constant for all frequencies) to represent earthquake ground motion. Actually, the simplest workable form of $G(\omega)$ is that corresponding to a band-limited white noise. The spectral density is constant in the frequency range from 0 to ω_o, as shown in Fig.3.4(a).

$$G(\omega) \begin{cases} = G_o & 0 \leq \omega \leq \omega_o \\ = 0 & \omega > \omega_o \end{cases} \qquad (3.2)$$

Based on Kanai's study[42] of the frequency content of a limited number of recorded strong ground motions. Tajimi[43] suggested the following widely quoted form for the spectral density function of ground motion (fig.3.4(b)):

$$G(\omega) = \frac{\left[1 + 4\zeta_g^2 (\omega/\omega_g)^2\right] G_o}{\left[1 - (\omega/\omega_g)^2\right]^2 + 4\zeta_g^2 (\omega/\omega_g)^2} \qquad (3.3)^{*)}$$

Sample functions x(t) with spectral densities corresponding to Eq.3.3 can be obtained by filtering "ideal white noise" (i.e., with $\omega_o = \infty$ in Eq.3.2) through a simple oscillator with natural frequency ω_g and viscous damping ζ_g[31]. These may be interpreted as the "predominant ground frequency" and the "ground damping", respectively. The values $\omega_g = 4\pi$ and $\zeta_g = 0.60$ have been suggested for firm ground sites. G_o is a measure of ground intensity.

Extensions of Eq.3.3 have been proposed to model ground motions whose spectral density shows more than one dominant spectral peak[46]. Ground motion models which account for the time-varying nature of the relative frequency content have

*) Soviet researchers[44,45] suggested a probabilistic approach to earthquake engineering based on the autocorrelation function of a form corresponding to the s.d.f. in Eq.3.3.

also been proposed[47]. In reference to Eq.3.1, it suffices to allow the amplitudes A_i to vary slowly with time. The spectral content can then be described by an "evolutionary" spectral density function $G(\omega,t)$, i.e., $G(\omega_i,t)$ is proportional to A_i^2 [48]. However, time-invariant models in which $G(\omega)$ reflects the frequency content during the most intense part of the ground motion are believed to be sufficiently accurate for the purpose of seismic response prediction, for all but certain nonlinear systems.

The integral over frequency of $G(\omega)$ equals the average total power, or the variance σ^2 for motions which fluctuate about a zero mean value, e.g., ground acceleration and linear system response. For the band-limited noise motion (Eq.3.2) the variance is:

$$\sigma^2 = \int_0^\infty G(\omega)d\omega = G_o\omega_o \tag{3.4}$$

and for the Kanai-Tajimi spectrum (Eq.3.3), the variance is:

$$\sigma^2 = \int_0^\infty G(\omega)d\omega = \frac{\pi G_o\omega_g}{4\xi_g}(1 + 4\xi_g^2)$$

Actually, a more useful way for dealing with the frequency content of ground motions is through the normalized spectral density function:

$$G^*(\omega) = \frac{1}{\sigma^2} G(\omega) \tag{3.6}$$

or its cumulative spectral distribution:

$$F^*(\omega) = \int_0^{\omega} G^*(\omega)d\omega \qquad (3.7)$$

which increases from 0 to 1 as ω goes from 0 to ∞. Note the analogy between normalized s.d.f. and the probability density function (p.d.f.) of any random variable: both are nonnegative and have unit area. The moments of the spectral density function $G(\omega)$ are:

$$\lambda_i = \int_0^{\infty} \omega^i\, G(\omega)d\omega = \sigma^2 \int_0^{\infty} \omega^i\, G^*(\omega)d\omega = \sigma^2\, \lambda_i^* \qquad (3.8)$$

in which λ_i^* is the ith moment of the unit area spectral density. It is clear that $\sigma^2 = \lambda_o$ and $\lambda_o^* = 1$. A measure of where the spectral mass is concentrated along the frequency axis is (see Fig.3.1):

$$\Omega = \sqrt{\lambda_2/\lambda_o} = \sqrt{\lambda_2^*} \qquad (3.9)$$

which is analogous to the root-mean-square (r.m.s.) of a random variable. A convenient measure of the spread or the dispersion of the s.d.f. about its center frequency is[49,50]:

$$\delta = \sqrt{1 - \lambda_1^2/\lambda_o\lambda_2} = \sqrt{1 - \lambda_1^{*2}/\lambda_2^*} \qquad (3.10)$$

which is dimensionless, always lies between 0 and 1, and increases with increasing bandwidth. Pursuing the analogy between $G^*(\omega)$ and the p.d.f. of a random variable, δ is equivalent to the ratio of the standard deviation to the root-mean-square value (see Fig.3.1). It is clear that δ will be large if $G(\omega)$ has two or more fairly widely separated peaks. Important time domain interpretations of the spectral parameters Ω and δ are discussed subsequently. In the frequency domain, these two parameters provide a summary description of $G^*(\omega)$. In fact, it is possible to develop (Chebychev-type) bounds on $F^*(\omega)$ in terms of Ω and δ [50+)]. A parameter whose definition is similar to δ has been proposed by Longuet-Higgins[51]:

$$\varepsilon = \sqrt{1 - \lambda_2^2/\lambda_o\lambda_4} = \sqrt{1 - \lambda_2^{*2}/\lambda_4^*} \qquad (3.11)$$

It also varies between 0 and 1, and it is closely related to the kurtosis parameter for random variables.

In evaluating higher spectral moments, problems due to lack of convergence are sometimes encountered. These are due to the fact that various proposed algebraic expressions (e.g., Eq.3.3), while providing a good fit to computed power spectra in the central frequency region, do not properly represent the ground motion in the upper frequency tail. Owing to recording and processing limitations, strong motion accelerograms provide little information about motion frequency content beyond a circular frequency $\omega_o = 2\pi/\Delta T$,

+) One such inequality is:

$$1 - F^*(\omega) \leq \delta^2 \left[(\omega/\Omega) - 1\right]^{-2}$$

For example, taking $\delta = 0.2$ and $\omega = 3\Omega$ yields the inequality $1 - F^*(3\Omega) \leq 0.01$; in words, for processes having a δ-factor equal to 0.2, the fraction of the total power contributed by components with frequencies larger than 3Ω is less than 1 %.

where T is approximately 0.02 seconds. Consequently, it is difficult to evaluate contributions to, say, the second spectral moment, due to frequencies beyond ω_o; one is, in effect, restricted to computing partial moments of $G(\omega)$, in a limited frequency range $(0, \omega_o)$.

The r.m.s. value σ is closely tied to the maximum ground acceleration, A both of which are more difficult to predict than the spectral parameters Ω and δ. The relation between σ and A is examined in some detail in Section 5. In particular, it is shown there that the median maximum acceleration, $\hat{a}$, is proportional to σ:

$$\hat{a} = \sigma \times \sqrt{2 \ln \quad 2.8 \frac{\Omega s}{2\pi}} \qquad (3.12)$$

in which s = strong-motion duration, and $(\Omega s)/2\pi$ = expected number of cycles of ground motion. The difficulty in the direct assessment of σ^2 from a ground motion record stems from the fact that the motion intensity actually varies with time and that the choice of the length of the motion is critical in computations of average power. In modeling ground motions for the purpose of random vibration analysis, a reasonable procedure is to relate σ directly to $\hat{a}$ (using Eq.3.12), to choose an equivalent duration of strong shaking, and to determine the spectral shape function $G^*(\omega)$, either directly from the response spectrum S_V or by selecting a theoretical shape (e.g., the Kanai-Tajimi form Eq.3.3).

It is of much interest to examine the relationship between the common characterizations of the earthquake threat at a site, i.e., response spectra, representative (real or artificial) ground motions and spectral density functions. The question of compatibility between these three representations is explored in the next two sections which outline the basic methodology of random vibration to predict the peak response of simple one-degree systems. From this, a procedure can be developed to generate artificial earthquake motions whose computed response spectra "match" a set of prescribed smooth response spectra (for details, see[1]).

4. THE VARIANCE OF THE SEISMIC RESPONSE OF LINEAR SYSTEMS

Response Variance for One-Degree Systems

Steady-State Response

A basic result of stationary random vibration of linear systems is the following relationship between the spectral density functions (s.d.f.) of input and output (see, for example[36]):

$$G_y(\omega) = G(\omega)\ |H(\omega)|^2 \tag{4.1}$$

in which $G_y(\omega)$ = the output s.d.f., $G(\omega)$ = the input s.d.f., and $|H(\omega)|$ = the amplification function of the linear system, i.e., the amplitude of the steady-state response of the system to a sinusoidal input with unit amplitude and frequency ω. Also, the variance of the response, σ_y^2, is equal to the area under $G_y(\omega)$:

$$\sigma_y^2 = \int_0^\infty G_y(\omega)d\omega = \int_0^\infty G(\omega)\ |H(\omega)|^2\ d\omega \tag{4.2}$$

For a linear one-degree system with natural frequency ω_n and damping ratio ζ, whose input and output are support acceleration and relative displacement response, respectively, the squared amplification function is:

$$|H(\omega)|^2 = \left[(\omega_n^2 - \omega^2) + 4\zeta^2\ \omega_n^2\ \omega^2\right]^{-1} \tag{4.3}$$

The relative displacement response variance σ_y is obtained by inserting Eq.4.3 into Eq.4.2. Also, the standard deviation of the pseudo-acceleration response is $\sigma_a = \omega_n^2\ \sigma_y$.

Note that, at the extremes of the frequency scale, this formulation leads to the following desirable, not commonly recognized, results. First, σ_a approaches the standard deviation of the ground acceleration (σ) when $\omega_n \rightarrow \infty$

(and $|H(\omega)|^2 - \omega_n^{-4}$):

$$\sigma_a = \omega_n^2 \sigma_y \rightarrow \left[\int_0^\infty G(\omega)d\omega\right]^{1/2} = \sigma \qquad (4.4\text{-a})$$

At the other extreme, when $\omega_n \rightarrow 0$ and $|H(\omega)| \rightarrow \omega^{-2}$ (see Eq.4.3), we obtain:

$$\sigma_y^2 = \int_0^\infty \omega^{-4} G(\omega)d\omega = \sigma^2_{\text{gr.displ.}} \qquad (4.4\text{-b})$$

The integrand $\omega^{-4}G(\omega)$ is actually the spectral density function of the ground displacement*, and therefore σ_y^2 is equal to the variance of the ground displacement when $\omega_n = 0$.

As shown in Fig.4.1, the earthquake excitation spectral density functions often varies relatively smoothly in the immediate vicinity of the system's natural frequency ω_n, while $|H(\omega)|^2$ exhibits a sharp peak at ω_n. This effect is more pronounced as the system damping decreases. It leads to the following useful approximation for the pseudo-acceleration response variance, $\sigma_a^2 = \omega_n^4 \sigma_y^2$:

$$\sigma_a^2 = \omega_n^4 \int_0^\infty G(\omega)|H(\omega)|^2 d\omega \simeq \omega_n^4 G(\omega_n) \int_0^\infty |H(\omega)|^2 d\omega - \omega_n G(\omega_n) +$$

$$+ \int_0^{\omega_n} G(\omega)d\omega = G(\omega_n)\omega_n \left[\frac{\pi}{4\xi} - 1\right] + \int_0^{\omega_n} G(\omega)d\omega \qquad (4.5)$$

Fig.4.1 illustrates the meaning of the two terms on the

*) This follows from Eq.4.1 and from the fact that the amplification function of a hypothetical system whose input is the second derivative of the output, equals ω^{-2}. Of course, in this section, both ground acceleration and ground displacement are assumed to be stationary random processes.

right side of Eq.4.5. The first term accounts for the contribution in a narrow frequency range around ω_n, the second the contribution in frequency range below $(0,\omega_n)$. The term $\omega_n G(\omega_n)$ is subtracted because the area it represents would otherwise be counted twice. The relative importance of the second term in Eq.4.5, the partial area under $G(\omega)$, increases for higher natural frequencies. For lightly damped systems with intermediate natural frequencies, the first term will strongly dominate. Eq.4.5 correctly predicts σ_a^2 when the excitation is an ideal white noise, i.e., $G(\omega) = G_o$, for all ω. We have then:

$$\sigma^2 = G_o\ \omega_n \left[(\pi/4\xi) - 1\right] + G_o\ \omega_n = (\pi\ G_o\ \omega_n)/4\xi \qquad (4.6)$$

A widely used approximation for the variance of the response to wide band excitation is obtained by substituting G_o in Eq.4.6 by $G(\omega_n)$. It is evident that this result is of little use in predicting responses at higher frequencies.

It is convenient to express the response variance σ_a^2 in terms of the ground motion variance σ^2, i.e., to evaluate the ratio :*)

$$F_a = \frac{\sigma_a^2}{\sigma^2} \simeq G^*(\omega_n)\omega_n \left[\frac{\pi}{4\xi} - 1\right] + F^*(\omega_n) \qquad (4.7)$$

*) For light damping, the factor $\left[(\pi/4\xi) - 1\right]$ may be replaced by $\pi/4\xi$.

where $G^*(\omega_n)$ = the unit-area ground motion s.d.f., and $F^*(\omega_n)$ = the normalized cumulative spectrum. It is not important that the expression for $F^*(\omega_n)$ in Eq.4.7 be accurate in the range where the first term predominates. But $F^*(\omega)$ should approach 1 at high values of ω_n, when the contribution due to the first term vanishes. If $F^*(\omega)$ is not easily obtainable, a convenient approximation which satisfies this criterion is the Rayleigh cumulative function:

$$F^*(\omega_n) = 1 - e^{-\omega_n^2/\Omega^2} \tag{4.8}$$

implying a.s.d.f. shape which has the correct mean-square frequency Ω^2 and an exponentially decaying upper tail. The fact that spectra such as the Kanai-Tajimi form poorly represents the very high frequency content of actual ground motions further justifies the use of approximations as Eq.4.8.

Transient Response

The steady-state value given by Eq.4.5 may never be closely approached when the one-degree system's natural frequency or damping are very small, i.e., when the product $\zeta\omega_n$ is small. As was first shown by Caughey and Stumpf[52] the response variance will build up from zero (at the time when the earthquake strikes) to a maximum value, near the end of the (equivalent stationary) motion duration, s. The frequency content of the one-degree system response will evolve in a way which can perhaps most conveniently be described by the time-dependent spectral density function $G_y(\omega,t)$[53]. For a broad class of functions $G_y(\omega,t)$, the time-dependent variance of the transient response can be obtained by integration over all frequencies:

$$\sigma_y^2(t) = \int_0^\infty G_y(\omega,t)\,d\omega \qquad (4.9)$$

The function $G_y(\omega,t)$ will depend on the input s.d.f. $G(\omega)$ and on the system properties. For any linear system with impulse response function $h(t)$, it is possible to define the truncated Fourier transformation, or the "time-dependent transfer function", as follows

$$H(\omega,t) = \int_0^t h(\tau)e^{-i\omega\tau}d\tau \qquad (4.10)$$

which converges to the transfer function $H(\omega)$ when $t \to \infty$. When the system is suddenly exposed to a steady excitation with spectral density function $G(\omega)$, the time-dependent response s.d.f. will be given by

$$G_y(\omega,t) = G(\omega)\,\left|H(\omega,t)\right|^2 \qquad (4.11)$$

In the case at hand, $G(\omega)$ is wide-band and smoothly varying and the system is a simple oscillator with impulse response function

$$h(t) = \begin{cases} \dfrac{1}{\omega_1} e^{-\xi\omega_n t} \sin \omega_1 t & t \geq 0 \\ 0 & t < 0 \end{cases} \qquad (4.12)$$

where $\omega_1 = \omega_n \sqrt{1 - \xi^2}$ = the damped natural frequency of the system. The transient squared amplification function $|H(\omega, t)|^2$ has the following form[52,54,55]:

$$|H(\omega,t)|^2 = |H(\omega)|^2 \Big\{1-2e^{-\omega_n \xi t}\Big[(\cos \omega_1 t+\frac{\omega_n \xi}{\omega_1}\sin \omega_1 t)\cos \omega t$$

$$+ \frac{\omega}{\omega_1} \sin \omega_1 t \sin \omega t - e^{-\omega_n \xi t}(\frac{1}{2} + \frac{\omega_n \xi}{\omega_1} \sin \omega_1 t \cos \omega_1 t$$

$$+ \frac{(\omega_n \xi)^2 - \omega_1^2 + \omega^2}{2\omega_1^2} \sin^2\omega_1 t)\Big]\Big\} \qquad (4.13)$$

Integrating this expression over all frequencies yields approximately

$$\int_0^\infty |H(\omega,t)|^2 \, d\omega \simeq \frac{\pi}{4\xi\omega_n^3} (1 - e^{-2\xi\omega_n t}) \qquad (4.14)$$

The foregoing integral increases from 0 to the stationary value $\pi/(4\xi\omega_n^3)$, which will be achieved when $t \gg 1/\xi\omega_n$. Comparison between Eq.4.14 and the stationary value motivates the definition of a fictitious <u>time-dependent damping</u>:

$$\xi_t = \frac{\xi}{1 - e^{-2\xi\omega_n t}} \qquad (4.15)$$

so that the right side of Eq.4.14 can be written as $\pi/4\zeta_t\omega_n^3$. Of course, $\zeta_t = \zeta$ when $t \to \infty$. Actually, the parameter ζ_t is particularly useful in that it allows the entire set of spectral shapes $|H(\omega,t)|^2$ (Eq.4.14) to be crudely approximated by:

$$|H(\omega,t)|^2 \simeq \left[(\omega_n^2 - \omega^2)^2 + 4\zeta_t\omega_n^2\omega^2\right]^{-1} \qquad (4.16)$$

which has the familiar form of the squared amplification function $|H(\omega)|^2$ given by Eq.4.3. The damping parameter ζ_t decays from a very high value down to the actual system damping ζ; the rate of decay is governed by the product $\zeta\omega_n$. The approximate form yields, for all values of t, not only about the same total area, but also the same central frequency (ω_n) and about the same "bandwidth" as the exact form. The main advantage resulting from the use of Eq.4.16 is that all the stationary results obtained in the previous section can now be applied to the transient response situation, simply by substituting ζ by ζ_t.

Note that the use of Eqs.4.15 and 4.16 also provides a convenient way for treating the response of the undamped system, for which the stationary condition is, of course, never closely approached. In this case

$$\zeta_t = \frac{\zeta}{1 - e^{-2\zeta\omega_n t}} \xrightarrow[\zeta=0]{} \frac{1}{2\omega_n t} \qquad (4.17)$$

A result of particular importance is the pseudo-acceleration response variance evaluated at the end of the motion duration, s:

$$\sigma_a^2(s) = \omega_n^4 \sigma_y^2(s) = \omega_n^4 \int_0^\infty G(\omega)|H(\omega,s)|^2 d\omega$$

$$\simeq G(\omega_n)\omega_n \left[\frac{\pi}{4\xi_s} - 1\right] + \int_0^{\omega_n} G(\omega)d\omega \qquad (4.18)$$

Also, the ratio of the transient acceleration response variance is:

$$F_a^2 = \frac{\sigma_a^2(s)}{\sigma^2} \simeq G^*(\omega_n)\omega_n \left[\frac{\pi}{4\xi_s} - 1\right] + F^*(\omega_n) \qquad (4.19)$$

These results are analogous in form to Equations 4.5 and 4.7 and, in fact, converge to them when the product $(\xi\omega_n s)$ grows large. Again, if the damping is light, little accuracy is lost by replacing the factor $[(\pi/4\xi_s) - 1]$ by $\pi/4\xi_s$ in Eqs.4.18 and 4.19.

Response Variance for Multi-Degree Systems

The basic procedure is identical to that followed for one-degree systems. The system relating input acceleration x(t) and output relative displacement y(t) has the impulse response function (see Eq.2.1):

$$h(t) = \sum_{k=1}^{n} c_k\, h_k(t) \qquad (4.20)$$

where $h_k(t)$ is the impulse response function of a one-degree system (see Eq.4.12) with parameters ω_k and ζ_k. The truncated Fourier transform of h(t), or the time-dependent transfer function, is:

$$H(\omega,t) = \int_0^t h(\tau)e^{-i\omega\tau}d\tau = \sum_{k=1}^{n} c_k H_k(\omega,t) \qquad (4.21)$$

and the time-dependent spectral density function of y(t) equals:

$$G_y(\omega,t) = G(\omega)\,|H(\omega,t)|^2 =$$

$$G(\omega)\sum_{k=1}^{n}\sum_{j=1}^{n} c_k\, c_j\, H_k(\omega,t)H_j^*(\omega,t) \qquad (4.22)$$

in which $H_j^*(\omega,t)$ = the complex conjugate of $H_j(\omega,t)$. Although the expression $G_y(\omega,t)$ has the form of a double summation, significant contributions to its moments with respect to frequency (spectral moments) usually come from the terms for which j = k. This is particularly true when modal frequencies are well-separated and when damping values are low. Algebraic manipulation allows one to express Eq.4.22 approximately as follows[50*)]:

$$G_y(\omega,t) \quad G(\omega)\sum_{k=1}^{n}|H_k(\omega,t)|^2\left\{c_k + \sum_{j\neq k} c_j\, c_k\, A_{kjt}\right\} \qquad (4.23)$$

*) The use of Eq.4.23 to evaluate the variance σ_y^2 results in a percentage error which is of the order of the square of the damping factor.

in which A_{kjt} is a factor which depends on the ration of modal frequencies $r = \omega_j/\omega_k$ and on the equivalent damping values $\xi_{kt} = \xi_k(1 - e^{-2\xi_k\omega_k t})^{-1}$ and

$$\xi_{jt} = \xi_j(1 - e^{-2\xi_j\omega_j t})^{-1} .$$

$$A_{kjt}=\frac{8r\,\xi_{kt}(\xi_{jt}+\xi_{kt}\,r)\left[(1-r^2)^2-4r(\xi_{kt}-\xi_{jt}\,r)(\xi_{jt}-\xi_{kt}r)\right]}{8r^2\left[(\xi_{kt}^2+\xi_{jt}^2)(1-r^2)^2-2(\xi_{jt}^2-\xi_{kt}^2r^2)(\xi_{kt}^2-\xi_{jt}^2r^2)\right]+(1-r^2)^4} \tag{4.24}$$

The factor A_{kjt} is plotted in Figure 4.2 as a function of r for different pairs of values of ξ_{jt} and ξ_{kt}. At r = 1, $A_{kjt} = 2\xi_{kt}/(\xi_{kt} + \xi_{jt})$, which is equal to one if $\xi_{jt} = \xi_{kt}$ *). A_{kjt} vanishes when r is either very small or very large. Integrating Eq.4.23 over all frequencies gives the time-dependent variance of the multi-degree system response y(t):

$$\sigma_y^2(t) = \int_0^\infty G_y(\omega,t)\,d\omega \simeq \sum_{k=1}^{n}(c_k^2 + \sum_{j\neq k} c_j\, c_k\, A_{kjt})\, \sigma_{y_k}^2(t) =$$

$$\sum_{k=1}^{n} \alpha_{kt}\, c_k^2\, \sigma_{y_k}^2(t) \tag{4.25}$$

where

$$\alpha_{kt} = 1 + \sum_{j\neq k}(c_j/c_k)\, A_{kjt} \tag{4.26}$$

---

*) Also, $A_{kjt} = r(\xi_{jt}/\xi_{kt})A_{kjt}$, so that $A_{jkt} + A_{kjt} = 2$ if r = 1.

It is easy to see that the "cross-terms" will be insignificant, i.e., $\alpha_{kt} \simeq 1$, when (i) the modal frequencies are well-separated, (ii) the damping values are small, and (iii) time t is sufficiently large.

The root-sum-square rule for combining modal maxima (Eq.2.2) rests on the assumption of statistical independence among modes. It applies theoretically to the standard-deviations of the modal contributions, as can be seen by putting $\alpha_{kt} = 1$ in Eq.4.25. Note that Eq.4.25 can be converted into an improved rule for combining modal maxima when modal interactions are important (i.e., by replacing $\sigma_y(t)$ by S and $\sigma_{y_k}(t)$ by S_{dk} in Eq.4.25).

5. PEAK FACTORS AND OTHER RESPONSE STATISTICS

Spectral Moments and Level Crossing Statistics

The distribution of the maximum response, as well as some other useful response "level crossing" statistics, depends importantly on the higher moments of the response spectral density function. If the response is stationary, its spectral moments can be obtained in the same way as those of the ground motion (see Eq.3.7).

$$\lambda_{i,y} = \int_0^{\infty} \omega^i \, G_y(\omega)\,d\omega \tag{5.1}$$

Throughout this section, the subscript y is used to characterize response parameters. When i = 0, Eq.5.1 defines the variance, i.e., $\lambda_{0,y} = \sigma_y^2$. It is well-known that $\lambda_{2,y} = \sigma_{\dot{y}}^2$ = the variance of the derivative of y(t). The center frequency is:

$$\Omega_y = \sqrt{\lambda_{2,y}/\lambda_{0,y}} = \sigma_{\dot{y}}/\sigma_y \tag{5.2}$$

For Gaussian processes, Ω_y is closely related to ν_a, the average number of times per second the response y(t) exceeds the response level a, or the average rate of a-upcrossings[56]

$$\nu_a = (\Omega_y/2\pi)\exp\left\{-a^2/2\sigma^2\right\} = \nu_o \exp\left\{-r^2/2\right\} \tag{5.3}$$

in which $r = (a/\sigma_y)$ and $\nu_o = \Omega_y/2\pi$. Note that for a = 0, we obtain $\nu_o = \Omega_y/2\pi$. Also, the average number of times per second that y(t) moves outside the range (-a,a) is $2\nu_a$. Another useful spectral parameter is:

$$\delta_y = \sqrt{1 - \lambda_{1,y}^2/\lambda_{0,y}\,\lambda_{2,y}} \tag{5.4}$$

which measures the spread or the variability in the frequency content of the response motion; it is dimensionless and lies between 0 and 1. The value of δ_y is small for narrow-band processes (it equals zero for a pure sinusoid with random phase angle) and relatively large for wide-band processes. In the time domain, δ_y is equal to the ratio of $\sigma_{\dot{r}}/\sigma_{\dot{y}}$, in which $\sigma_{\dot{r}}$ is the r.m.s. value of the slope of the envelope r(t) of the function y(t), and $\sigma_{\dot{y}}$ is the r.m.s. of the slope y(t). The envelope definition used here is that of Cramer and Leadbetter[57]. It exists for any stationary random process regardless of bandwidth. Of course, the concept of an envelope is most useful for narrow-band random functions. The mean number of times per second the envelope of y(t) exceeds the level a, or the mean rate of envelope crossings, is given by[57,50]:

$$n_a = \sqrt{2\pi}\;\delta_y \left[\frac{a}{\sigma_y}\right] \nu_a \tag{5.5}$$

The relationship between the mean crossing rates $2\nu_a$ and n_a is interesting. At low threshold levels, n_a will be less than $2\nu_a$: the crossings of y(t) outside the range (-a,a) tend to occur in clumps (see Fig.5.1) which follow envelope crossings. Lyon[58] has argued that the quotient $2\nu_a/n_a$ can be interpreted as the mean clump size. This concept is particularly useful when the ratio is well above 1. But for high thresholds, the ratio $2\nu_a/n_a$ may become much smaller than one, as many envelope crossings are not followed by a-crossings. An estimate of the mean clump size which accounts for this effect is [49,64]:

$$E\left[N_a\right] = \frac{1}{1 - \exp(-n_a/2\nu_a)} = \frac{1}{1 - \exp\{-\sqrt{\pi/2}\ \delta_y(a/\sigma_y)\}} \tag{5.6}$$

which is plotted in Fig.5.2 for several values of δ_y. Note that $E\left[N_a\right] \rightarrow 1.$ for high threshold levels. When the mean clump size is large, i.e., when the product $\delta_y(a/\sigma_y)$ is small, $E\left[N_a\right] \simeq 2\nu_a/n_a = \left[\sqrt{\pi/2}\ \delta_y(a/\sigma_y)\right]^{-1}$.

These definitions can easily be extended to the non-stationary response situation when the frequency content of y(t) can be described in terms of a time-dependent (or evolutionary) spectral density function $G_y(\omega,t)$. The time-dependent moments $\lambda_{i,y}(t)$ can then be obtained by integration of $\omega^i G(\omega,t)$, over all frequencies, i.e.,

$$\lambda_{i,y}(t) = \int_0^\infty \omega^i\ G(\omega)\ \left|H(\omega,t)\right|^2 d\omega \tag{5.7}$$

Furthermore, time-dependent response statistics, e.g., $\Omega_y(t)$ and $\delta_y(t)$, can be obtained by substituting the appropriate time-dependent moments in the "stationary case" definitions in much the same way as the variance $\sigma_y^2(t) = \lambda_{0,y}(t)$:

$$\lambda_{i,y}(t) \simeq \omega_n^{-4} \left\{G(\omega_n)\left[\omega_n^4 \int_0^\infty \omega^i \left|H(\omega,t)\right|^2\ d\omega - \frac{\omega_n^{i+1}}{i+1}\right] + \int_0^{\omega_n} \omega^i G(\omega)\, d\omega\right\} \tag{5.8}$$

In the case of one-degree systems, $|H(\omega,t)|^2$ may be substituted by the right side of Eq.4.16. This equation becomes identical to Eq.4.18 when i = 0. At intermediate and low natural frequencies, the second term in Eq.5.8 will be uniimportant. In this case, one obtains the central frequency $\Omega_y(t) \simeq \omega_n$ and the dispersion parameter $\delta_y(t) \simeq [(4/\pi)\zeta_t]^{1/2}$, where ζ_t is given by Eq.4.15.At very high natural frequencies, the second term will predominate in Eq.5.8 and the spectral parameters of the response approach those of the ground motion, i.e., $\Omega_y(t) \rightarrow \Omega$ and $\delta_y(t) \rightarrow \delta$.

In general, the spectral shape parameters $\Omega_y(t)$ and $\delta_y(t)$ will lie in between the values obtained for these limiting cases. The following formulas can then be used. They express the parameters of the spectral density function, say $G_T(\omega)$, which is the sum of a number of (positive) functions $G_k(\omega)$, each contributing a fraction p_k to the total area under $G_T(\omega)$. The spectral parameters of $G_T(\omega)$ are Ω_T and δ_T, those of $G_k(\omega)$ are Ω_k and δ_k. The weights p_j sum to one. We have:

$$\Omega_T = (\sum_k p_k \Omega_k^2)^{1/2} \tag{5.9}$$

$$\delta_T = \left\{1 - (\sum_k p_k \, (\Omega_k/\Omega_t)\sqrt{1 - \delta_k^2})\right\}^{1/2} \tag{5.10}$$

In the case at hand, there are two contributions with weights $p_1 = 1 - p_2$ and $p_2 = \sigma^2 F^*(\omega_n)/\sigma_y^2(s)$, and the parameters of interest are $\Omega_T = \Omega_y(t)$, $\delta_T = \delta_y(t)$, $\delta_1 = [(4/\pi)\xi_t]^{1/2}$, $\delta_2 = \delta$, $\Omega_1 = \omega_n$ and $\Omega_2 = \Omega$. The two limiting cases referred to earlier correspond to $p_1 = 0$ and $p_1 = 1$, respectively.

For multi-degree systems, higher spectral moments have the same form as Eq.4.25; the ith spectral moment is:

$$\lambda_{i,y}(t) = \int_0^\infty \omega^i G_y(\omega,t)d\omega \simeq \sum_{k=1}^{n} \alpha_{kt}\, c_k^2\, \lambda_{i,y_k}(t) \qquad (5.11)$$

where $\lambda_{i,y_k}(t)$ is the ith spectral moment for the kth mode (see Eq.4.34). Of course, $\lambda_{0,y_k}(t) = \sigma_y^2(t)$. Other spectral parameters and level crossing statistics can also be obtained for multi-degree system response.

Prediction of Peak Factors

The purpose of this section is to evaluate the factor $r_{s,p}$ by which the response standard deviation $\sigma_y(s)$ must be multiplied to predict the level $y_{s,p}$ below which the absolute value of the response y(t) will remain, with probability p, during the time interval (0,s). The task at hand is equivalent to finding the probability $L_a(s)$ that the system response fails to make a "passage" across a specified response level $\underline{a}$ during the time interval s. The first-passage problem has for decades been the subject of considerable research, and an exact solution does not yet exist. Moreover, few of the proposed approaches are of practical value to earthquake engineers*).

Most of the literature on the first-passage problem deals with the stationary response to Gaussian white noise of a lightly damped linear one-degree system. The reader is referred to Crandall[59] for an excellent state-of-the-art review. Below, a relatively simple approximate procedure is presented to predict the maximum responses of a rather general linear system exposed, suddenly and for a limited time, to steady-state Gaussian excitation. The proposed solution is based on research by the writer and his associates[49,58,55]. The case considered first is when the random motion at hand is stationary and Gaussian and the starting condition is random.

Stationary Response:

Theoretical and simulation studies[60,61] have confirmed that the propability $L_a(t)$ decays approximately

*) In fact, the most common approach to this problem has been to avoid it by adopting a constant peak factor, e.g., by predicting maximum response based on the "3σ rule".

exponentially with time, as follows*):

$$L_a(s) = A \exp\{-\alpha s\} \tag{5.12}$$

in which $L_a(s)$ = the probability that $|y(t)|$ remains below the level a during the interval $(0,s)$, $A = L_a(0)$ = the probability of starting below the threshold, and a = the decay rate. At high levels, $A \simeq 1$, and $\alpha \approx 2\nu_a$ (see Eq.5.3):

$$L_a(t) = \exp\{-2\nu_a t\} = \exp\{-(\Omega_y/\pi)\exp(-a^2/2\sigma_y^2)t\} \tag{5.13}$$

Cramer[62] has shown that this result is asymptotically exact (when the level a increases to infinity). A.G.Davenport[63] independently derived this result. It is consistent with the assumption that high level crossings occur according to a Poisson process. For a-levels of practical interest, however, the use of Eq.5.13 results in an error whose magnitude strongly depends on the bandwidth of the process. Numerical simulation studies indicate that the error tends to be on the unsafe side for wide-band processes and low threshold levels[61] and on the safe side for narrow-band processes[60]. For wide-band processes, the main effect is that the Poisson crossing assumption makes no allowance for the time the motion actually spends above the level a. For narrow-band processes, it is important to account for the fact that

*) The validity of the exponential approximation is doubtful at very small values of s, e.g., when only a few cycles of motion have elapsed. This is of little practical concern here.

evel crossings tend to occur not independently in accordance with the Poisson model, but in clusters or clumps.

The use in Eq.5.12 of the following approximation for α yields much improved estimates of $L_a(t)$ which are in close agreement with those obtained by simulation:

$$\alpha = 2\nu_a \frac{1 - \exp\left\{-\sqrt{\pi/2}\ \delta_e r\right\}}{1 - \exp\left\{-r^2/2\right\}} \tag{5.14}$$

in which $r = a/\sigma_y$ is the reduced level and $\delta_e = \delta_y^{1+b}$ is a bandwidth measure; b is a semi-empirical nonnegative constant estimated at about 0.2. Chosing $b = 0$ (or $\delta_e = \delta_y$) will result in slightly more conservative probability estimates (i.e., $r_{s,p}$ will be slightly higher). The ratio $\alpha/2\nu_a$ given by Eq.5.14 converges to one at high threshold levels in accordance with Cramer's results. An improved estimate of the probability A (in Eq.5.2)*) is $(1 - \exp\{-r^2/2\})$, which also approaches one at high levels. For the derivation of Eq.5.14, see Vanmarcke[64].

The approximation for $L_a(s)$ based on Eqs.5.2 and 5.14 depends on the motion parameters σ_y, Ω_y and δ_y, all of which are defined in terms of the first few spectral moments of y(t). The reduced level $r_{s,p}$ corresponding to reliability p and duration s is obtained by inverting $p = \exp\{-\alpha s\}$. The

*) A is interpreted here as the probability that the first peak of $|y(t)|$ (immediately after the start) will be below the level $\underline{a} = r\,\sigma_y$.

result can be expressed in terms of the factors $n = (\Omega_y s/2\pi)(-\log p)^{-1}$ and δ_e. When $p = e^{-1} = 0.368$, $(-\log p)^{-1} = 1$, and n equals the average number of cycles of response motion $(\Omega_y s/2\pi)$. The values of $(-\log p)^{-1}$ corresponding to p = 0.5, 0.9, and 0.99 are about 1.4, 10, and 100, respectively. Note that the factor $(-\log p)^{-1}$ may be substituted by $(1 - p)^{-1}$ when the reliability p is very close to one.

The reduced level $r_{s,p}$ is plotted in Fig.5.3 as a function of n for several values of δ_e. For large values of δ_e, the solution approaches the upper bound curve which constitudes the solution to Eq.5.13. The exact equation for the upper bound curve is:

$$r_{s,p} = \sqrt{2 \log 2n} \tag{5.15}$$

in which:

$$n = (\Omega_y s/2\pi)(-\log p)^{-1} \tag{5.16}$$

An approximate expression for the other curves in Fig.5.3 is:

$$r_{s,p} = \left\{2 \log\left[2n(1 - \exp\left\{- \delta_e \sqrt{\pi \log 2n}\right\})\right]\right\}^{1/2} \tag{5.17}$$

Transient Response:

A major of the solution just presented is that it can easily be extended to obtain first-passage probability estimates for transient response whose frequency content is described in terms of the time-dependent spectral density function $G_y(\omega,t)$. It is possible to evaluate the time-dependent decay rate $\alpha(t)$ in the expression[65]:

$$L_a(s) = \exp\left\{-\int_0^s \alpha(t)dt\right\} \quad (5.18)$$

which is a direct extension of Eq.5.12. In this case, $A = L_a(0) = 1$, since the response y(t) builds up from the rest. Recall that our aim is to determine the factor $r_{s,p}$ which must be multiplied by the standard deviation $\sigma_y(s)$ to predict maximum response fractiles. A direct but impractical approach is to substitute the parameters σ_y, Ω_y and δ_y in the expression for α (Eq.5.14) by their time-dependent equivalents, and to solve Eq.5.14 numerically for $r_{s,p} = a/\sigma_y(s)$. A much simpler approximate procedure is outlined below. As the response variance $\sigma_y^2(t)$ increases with time, from 0 to $\sigma_y^2(s)$, the failure rate of $\alpha(t)$ will increase much more rapidly, and the integral $\int_0^s \alpha(t)dt$ in Eq.5.14 will be dominated by contributions corresponding to values of t close to s. This motivates the introduction of an "equivalent stationary response" duration s_o such that:

$$L_a(s) = \exp\left\{-\int_0^s \alpha(t)dt\right\} = \exp\left\{-s_o\ \alpha(s)\right\} \qquad (5.19)$$

where, clearly, $s_o \leq s$. The ratio (s_o/s) can be roughly estimated from the ratio

$$m = \sigma^2(s)/\sigma^2(s/2) \qquad (5.20)$$

by the following equation*):

$$s_o/s \simeq \exp\left\{-2(m - 1)\right\} \qquad (5.21)$$

For lightly damped one-degree systems, $1 \leq m \leq 2$, with the upper bound applicable when $\omega_n \beta s \rightarrow 0$. When $m = 1$, $s_o = s$. The attractive feature of Eq.5.19 is, of course, that the stationary first-passage solution can again be used to derive $r_{s,p}$. It suffices to substitute the quantities $s\,\Omega_y$ and δ_y in Eqs.5.15 to 5.17 by $s_o\Omega_y(s)$ and $\delta_y(s)$, respectively.

A comprehensive survey of approaches to the first passage problem with reference to seismic response may be found in Dr.Grossmayer's lecture notes.

---

*) Eq.5.21 is obtained by approximating the area under $\alpha(t)$ (from 0 to s) in terms of the values $\alpha(s)$ and $\alpha(0.5s)$, which in turn depend principally on the respective transient variances. Note that m is the ratio of the transient variances at s and 0.5 s, respectively.

6. OVERALL SEISMIC SAFETY ANALYSIS

The purpose of seismic safety analysis of structures is to express, in quantitative terms, the probability of "failure" within a specified period of time. "Failure" is said to occur when the structural system (or a major structural component) reaches a "limit state", such as first yield, excessive deflection, excessive acceleration, exceedance of the ultimate strength, or collapse. It is widely recognized that the safety evaluation requires these two basic steps:

(i) Determine the probability of "failure" given that an earthquake of known ground motion intensity occurs (e.g., given the peak ground acceleration $\underline{a}$, or the r.m.s. ground motion intensity σ). This "conditional" probability depends on the structure and its design.

(ii) Determine the probability of occurrence of an earthquake of given ground motion intensity within the time period of interest. This probability depends on the site location and on the regional seismicity.

If the ground motion intensity is characterized by the peak acceleration $\underline{a}$, then the overall seismic safety evaluation proceeds as follows. For "failure" to occur, two events must happen: first, a ground motion with intensity $\underline{a}$ must occur, and secondly, this motion must cause "failure". All possible values of $\underline{a}$ must be considered. The probability of failure P_F may be expressed as follows:

$$P_F = \int_a P_{F|a}\, p_a \, da \simeq \sum_{a_i} P_{F|a_i}\, p_{a_i} \tag{6.1}$$

in which $P_{F|a}$ = the conditional probability of failure given the intensity $\underline{a}$; $(p_a \, da)$ = the probability of occurrence of a ground motion with intensity between $\underline{a}$ and $\underline{a}$ + da. The values a_1, a_2,.., a_1,... provide a suitable discretization of the continuous intensity parameter.

Section 1 of these notes has described methodology to obtain the seismic risk estimates p_a or p_{a_i}. Frequently, the output of a seismic risk analysis is a plot of *annual* probabilities of *exceeding* a given ground motion level, say a_d, as a function of a_d. From these plots, the probabilities p_{a_i} needed in Eq.6.1 can be obtained by standard techniques of probability analysis.

Most of these notes have been devoted to methodology to determine conditional probabilities such as $P_{F|a}$. In the conventional deterministic methods summarized in Section 2, only a "point estimate" of the seismic load effect (or the seismic structural response) is obtained. No attempt is made to assess the variability of the load effect. The predicted load effect is compared to the allowable load effect (or the available seismic resistance). If the (predicted) load effect exceeds the (presicted) resistance, "failure" is thought to be certain.

In reality, of course, neither the seismic load effect (given a) nor the seismic resistance are perfectly predictable. Therefore, it makes sense to express structural performance under earthquake excitation of known intensity in probabilistic terms, i.e., through the conditional probability of failure $P_{F|a}$ which increases from 0 to 1 as the seismic intensity a increases from very small to very high levels. The sources of uncertainty influencing $P_{F|a}$ are listed below:

(1) random phasing of ground motion sinusoidal component
(2) strong-motion duration s
(3) dominant "ground period" ω_g (in Eq.3.3)
(4) structure's dynamic characteristics
(5) resistance or allowable load effect

Section 3 has discussed descriptions of earthquake ground motion in which the strong-motion duration s and the spectral density function $G(\omega)$ and its parameters are prescribed. The randomness of the phase angles of the contributing sinusoids (ϕ_i in Eq.3.1) is the only explicitly recognized source of uncertainty in the ground motion model. Similarly, the load effect distributions, i.e., the random vibration solutions, developed in Sections 4 and 5, only reflect uncertainty attributable to the random phasing of the sinusoids in the ground motion. In other words, the random vibration approach yields the conditional seismic load effect distribution given the ground motion parameters (such as $\underline{s}$ and ω_g) and the structure's dynamic properties (such as ω_n and ζ). In principle, the unconditional seismic load effect distribution can be expressed as a weighted combination of conditional distributions. Each "weight" is the relative likelihood of encountering a specific combination of ground motion and structural parameters. Finally, to obtain $P_{F|a}$, load effect and resistance distributions may be combined in the way this is usually done in classical reliability analysis. The methodology of overall seismic safety analysis just outlined is currently under intensive study at M.I.T. Probabilistic information is being gathered about earthquake ground motion parameters, structural dynamic properties and structural component resistance levels. The methodology is now being implemented for linear elastic structures[66]. Research on the seismic safety of inelastic structures is still at the stage where workable approximate random vibration solutions are sought (see Gazetas[2]).

ACKNOWLEDGMENTS

The writer is grateful to the Directorate of the International Centre for Mechanical Sciences in Udine, Italy and to Professor H.Parkus of the Technical University of Vienna for the opportunity to lecture on the timely subject of earthquake engineering. The lecture series, which had been scheduled nearly a year in advance, was held just two months after a disastrous earthquake devastated the Fruili region of Northern Italy.

Support for the writer's research in earthquake engineering was provided by the United States National Science Foundation under Grants GK-4151, GK-26296 and ATA 74-06935.

REFERENCES

[1] Vanmarcke, E.H., Structural response to earthquakes, Chapter 8 in Seismic Risk and Engineering Decisions, C. Lomnitz and E.Rosenblueth, Eds., Elsevier Book Co., Amsterdam, 1976.

[2] Gazetas, G., Random vibration analysis of inelastic multi-degree-of-freedom systems subjected to earthquake ground motion, Ph.D. Thesis, M.I.T. Dept. of Civil Engineering, August 1976.

[3] Newmark, N.M. and Rosenblueth, E., Fundamentals of Earthquake Engineering for Buildings, Prentice-Hall, Inc., Englewood Cliffs, N.J., 1971.

[4] Newmark, N.M. and Hall, W.J., Seismic design criteria for nuclear reactor faculities, Proc.4th WCEE, Santiago, Chile, II-B4, 37, 1969.

[5] Gutenberg, B. and Richter, C., Earthquake magnitude, intensity and acceleration, Bull.Seism.Soc.Am., 46, 105, 1958.

[6] Housner, G.W., Probabilistic aspects of earthquakes, Proc. of the ASCE-EMD Specialty Conference on Probabilistic Concepts and Methods in Engineering, Nov. 1969.

[7] Rosenblueth, E. and Esteva, L., Spectra of earthquakes at moderate and large distances, Soc.Mex. de Ing. Sismica, II, 1, 1, Mexico, 1964 (in Spanish).

[8] Cornell, C.A., Engineering seismic risk analysis, Bull. Seism.Soc. Am., 58, 5, 1583, 1968.

[9] Kallberg, K., Seismic risk in Southern California, M.I.T. Dept. of Civil Engineering Research Report R69-31, June 1969.

[10] Esteva, L., Seismic risk, Chapter 6 in Seismic Risk and Engineering Decisions, C.Lomnitz and E.Rosenblueth, Eds., Elsevier Book Co., Amsterdam, 1976.

[11] Cornell, C.A. and Vanmarcke, E.H., Seismic risk for offshore structures, Proc. 7th Offshore Technology Conf., Houston, Texas, May 1975.

[12] Cornell, C.A. and Merz, H., Seismic risk analysis based

on a quadratic frequency law, Bull. Seism.Soc.Am., 63, 6, 1999, December 1973.

[13] Donovan, N. and Bornstein, A.E., A Review of Seismic Risk Applications, Proceedings of the 2nd International Conf. on Appl. of Stat. and Prob. in Soil and Structural Engrg., published by Deutsche Gesellschaft für Erd- und Grundbau, Aachen, Germany, 265, September 1975.

[14] Penzien, J. and Watabe, M., Characteristics of 3-dimensional earthquake ground motions, Earthquake Eng. and Struct. Dyn., 3, 365, 1975.

[15] Biggs, M., Introduction to Structural Dynamics, McGraw-Hill, New York, 1964.

[16] Hurty, W.C., and Rubinstein, M.F., Dynamics of Structures, Prentice-Hall, Englewood Cliffs, 1964.

[17] Clough, R.W. and Penzien, J., Dynamics of Structures, McGraw-Hill, New York, 1975.

[18] Caughey, T.K., Classical normal modes in damped linear systems, J. Appl.Mech., 27, 269, June 1960.

[19] Roesset, J.M., Whitman, R.V. and Dobry, R., Modal analysis with foundation interaction, J.Struct.Div., Proc. ASCE, 99, ST3, March 1973.

[20] Bielak, J., Modal analysis for building-soil interaction, J. Eng. Mech. Div., Proc. ASCE, 102, EM 5, 771, 1976.

[21] Sarrazin, M., Roesset, J.M. and Whitman, R.V. Dynamic soil-structure interaction, J.Struct. Div., Proc. ASCE, 98, ST7, July 1972.

[22] Rosenblueth, E., A basis for aseismic design, Doctoral Thesis, University of Illinois, Urbana, 1951.

[23] Rosenblueth, E. and Elorduy, J., Response of linear systems to certain transient disturbances, Proc. 4th WCEE, A1-185, January 1969.

[24] Seed, H.B. and Idriss, I.M., Influence of soil conditions on ground motions during earthquakes, J.Soil Mech. and Found. Div., Proc. ASCE, 95, SM1, 99, January 1969.

[25] Crandall, S.H., Nonlinearities in structural dynamics, Shock and Vibration Digest, October 1974.

[26] Goto, H. and Iemura, H., Earthquake response of single-degree-of-freedom hysteretic systems, Proc. 5th WCEE, Paper 266, Rome, Italy, 1973.

[27] Blume, J., Structural dynamics in earthquake resistant design, Transactions ASCE 125, 1088, 1960.

[28] Newmark, N.M. and Hall, W.J., A rational approach to seismic design standards for structures, Proc. 5th WCEE, Rome, Italy, 2266, June 1973.

[29] Veletsos, A.S., Maximum deformations of certain non-linear systems, Proc. 4th WCEE, Santiago, Chile, II-A4, 155, 1969.

[30] Husid, R., The effect of gravity on the collapse of yielding structures with earthquake excitation, Proc. 4th WCEE, II, A4, 31, Chile, 1969.

[31] Penzien, J. and Liu, S.C., Nondeterministic analysis of nonlinear structures subjected to earthquake excitations. Proc. 4th WCEE, Santiago, Chile, I-A1, 114, 1969.

[32] Clough, R.W., Earthquake response of structures. Ch. 1w

in Earthquake Engineering, edited by R,L. Wiegel, Prentice-Hall, Englewood Cliffs, N.J., 1970.

[33] Biggs, J.M. and Grace, P.H., Seismic response of buildings designed by code for different earthquake intensities, M.I.T. Dept. of Civil Eng.,Res. Rep. R73-7, January 1973.

[34] Housner, G.W., Properties of strong ground motion earthquakes, BSSA, 37, 19, 1947.

[35] Housner, G.W., Behaviour of structures during earthquakes. Proc. ASCE, 85, EM4, 109, 1959.

[36] Crandall, S.H. and Mark, W.D., Random Vibration in Mechanical Systems, Academic Press, New York, 1963.

[37] Lin, Y.K. Probabilistic Theory of Structural Dynamics. McGraw-Hill, New York, 1967.

[38] Hou, S., Earthquake simulation models and their applications, M.I.T. Dept. of Civil Engineering, Report R68-17, May 1968.

[39] Shinozuka, M., Digital simulation of ground accelerations, Proc. 5th WCEE, 2829, Rome, Italy,June 1973.

a dynamic system under random excitation, J.Appl. Mech. 28,563, 1961.

[53] Corotis, R. and Vanmarcke, E.H., On the time-dependent frequency content of oscillator frequency content, to appear in J. Eng.Mech. Div., Proc. ASCE, 1975.

[54] Hammond, S.K., On the response of single and multi-degree of freedom systems to nonstationary random excitations. J. Sound and Vibration, 7, 393, 1968.

[55] Corotis, R., Vanmarcke, E.H. and Cornell, C.A., First passage of nonstationary random processes, J.Eng. Mech. Div., Proc. ASCE, 98, EM2, 401, 1972.

[56] Rice, S.O., Mathematical analysis of random noise, Bell System Technical Journal, Part I: 23, 282, 1944; Part II: 24, 46, 1945.

[57] Cramer, H. and Leadbetter, M.R., Stationary and Related Stochastic Processes, Wiley, New York, 1967.

[58] Lyon, R.H., On the vibration statistics of a randomly excited hard-spring oscillator, J. Acous. Soc. of Am., Part I: 32, 716, 1960; Part II: 33, 1395, 1961.

[59] Crandall, S.H., First-Crossing probabilities of the linear oscillator, J. Sound and Vibration, 12, 285, 1970.

[60] Crandall, S.H., Chandiramani, K.L., and Cook, R.G., Some first passage problems in random vibrations, J. Appl.Mech., 33, 532, September 1966.

[61] Ditlevsen, O., Extremes and first passage times with application in Civil Engineering, Doctoral Thesis, Technical University of Denmark, Copenhagen, 1971.

[62] Cramer, H., On the intersections between the trajectories of a normal stationary stochastic process and a high level, Arkiv. Mat., 6, 337, 1966.

[63] Davenport, A.G., Note on the distribution of the largest value of a random function with application to gust loading, Proc. Inst. of Civ. Eng., 28, 187, June 1964.

[64] Vanmarcke, E.H., On the distribution of the first-passage time for normal stationary random processes, J. Appl. Mech., 42, 215, March 1975.

[65] Amin, M., Tsao, H.S. and Ang, A.H., Significance of nonstationarity of earthquake motions, Proc. 4th WCEE, Santiago, Chile, 97, 1969.

[66] Gasparini, D., On the safety provided by alternative seismic design methods, Ph.D. Thesis, M.I.T. Dept. of Civil Engineering, August 1976.

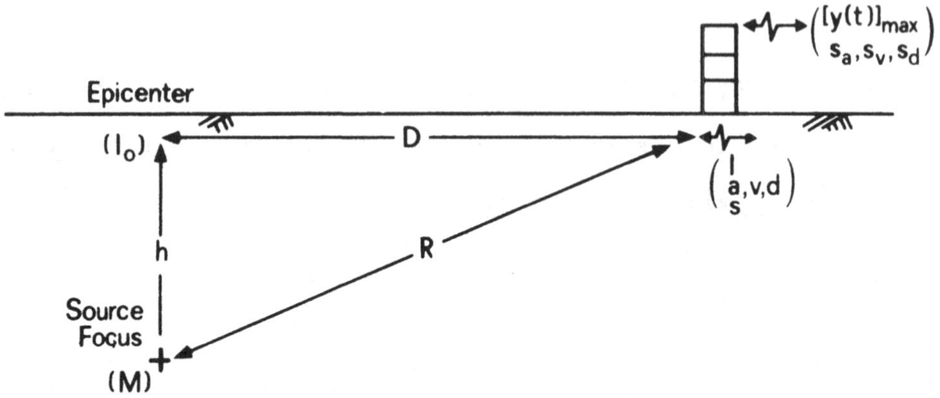

Fig. 1.1 Characteristics of the Earthquake, the Site Ground Motion and the Structural Response

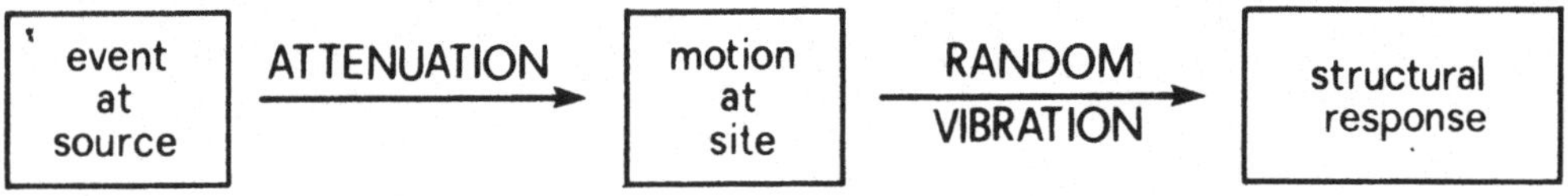

Fig. 1.2 Schematic Representation of Earthquake Effects

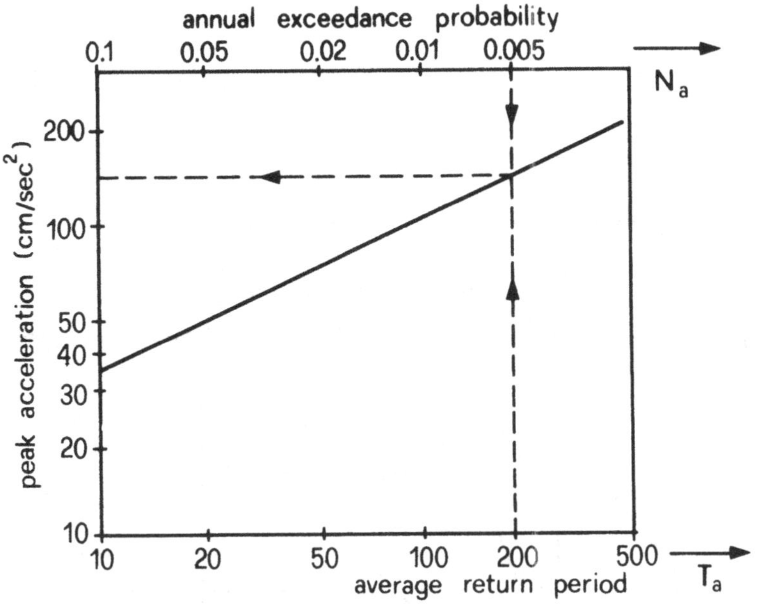

Fig. 1.3 Plot of Peak Ground Acceleration - versus - Average Return Period for a Site in Southern California

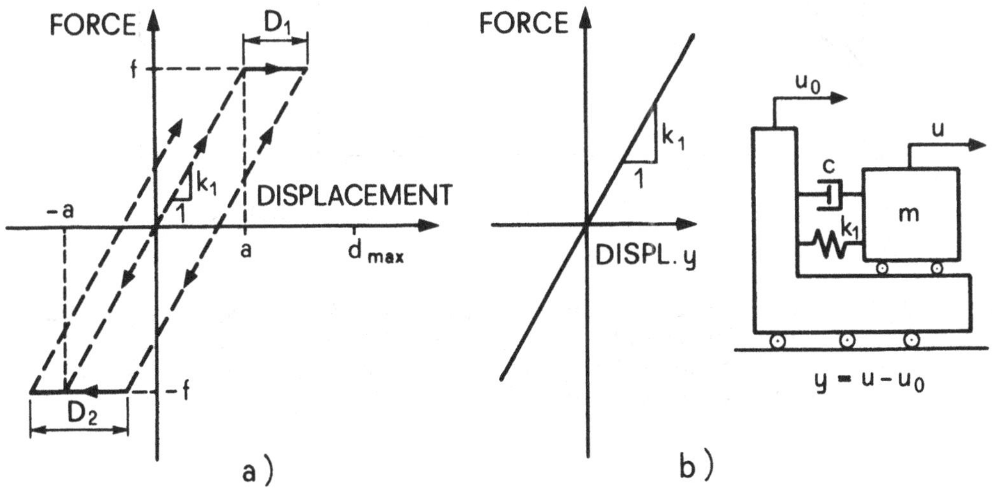

Fig. 2.1 (a) One-Degree Elasto-Plastic System and (b) Associated Linear System

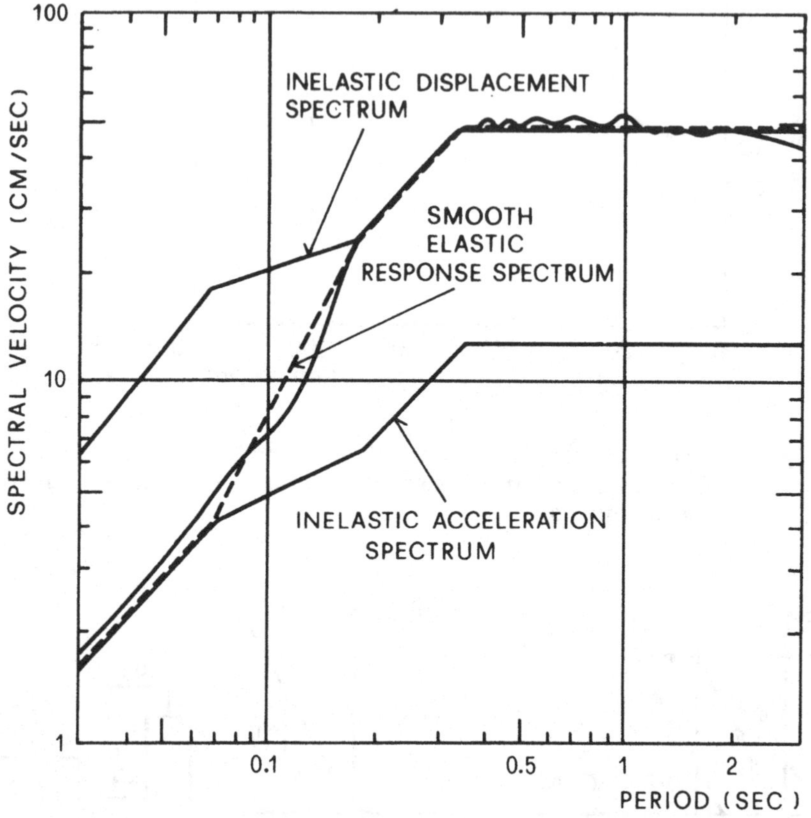

Fig. 2.2 Inelastic Response Spectrum for Damping $\zeta=0.02$ and Ductility Ratio $\mu=4$, Based on an Elastic 2 % Damped Response Spectrum (represented by the dotted line)

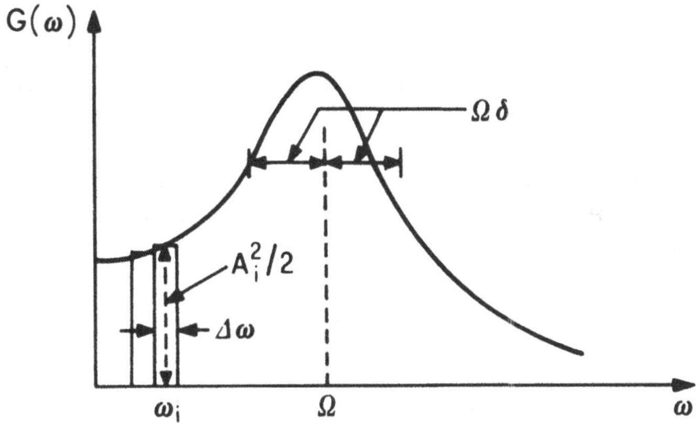

Fig. 3.1 The Spectral Density Function G(ω) and the Spectral Parameters Ω and δ.

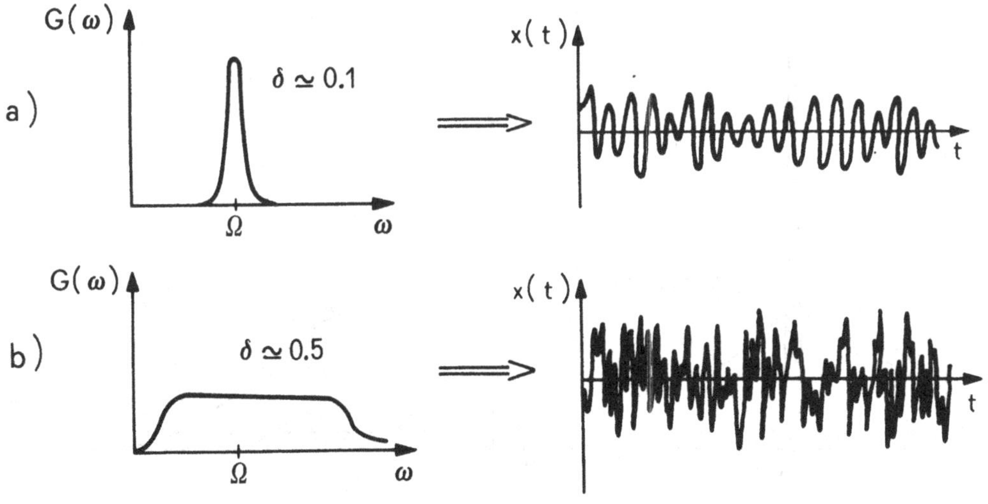

Fig. 3.2 Spectral Density Functions Corresponding to Different Bandwidths

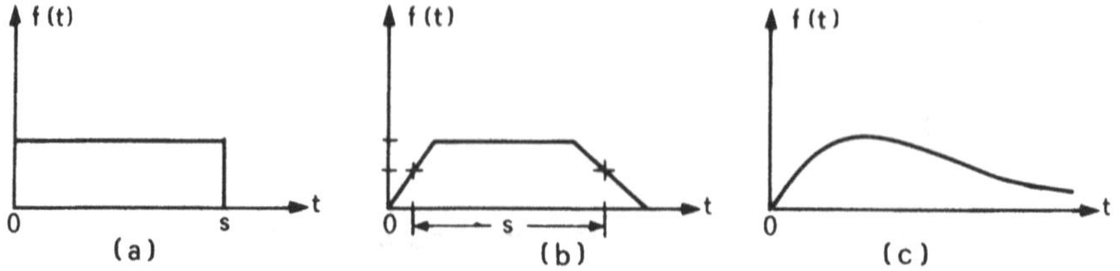

Fig. 3.3 Intensity Envelope Functions: (a) Boxcar; (b) Trapezoidal[38]; (c) Exponential[39].

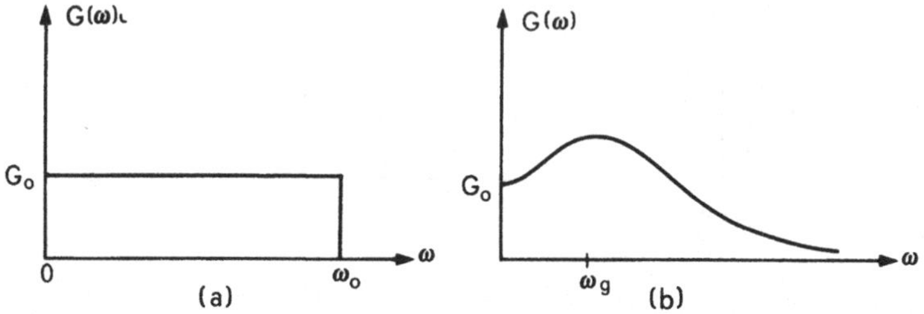

Fig. 3.4 (a) Band-Limited White Noise Spectral Density, and (b) Kanai-Tajimi Spectral Density

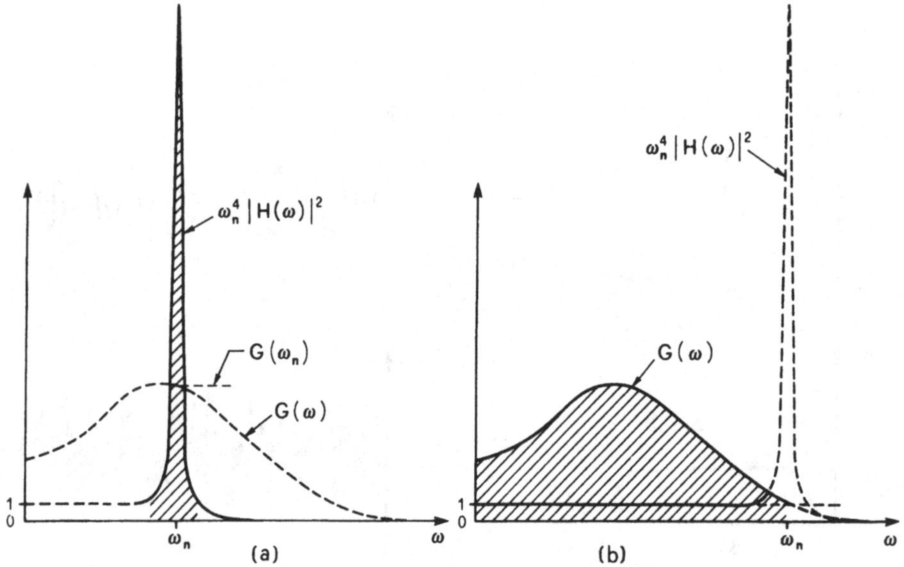

Fig. 4.1 Computation of the Variance of One-Degree System Response to Wide-Band Input. The first term (on the right side of Eq.4.5) accounts for the contribution in a narrow frequency range around ω_n, the second term for the contribution in the frequency range $(0, \omega_n)$.

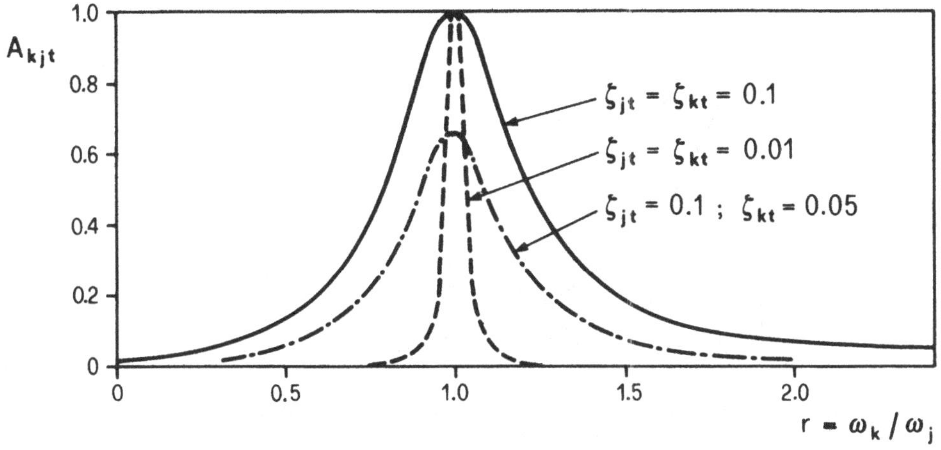

Fig. 4.2 The Factor A_{jkt} which Accounts for the Interaction Between Modes j and k of a Multi-Degree-Of-Freedom System

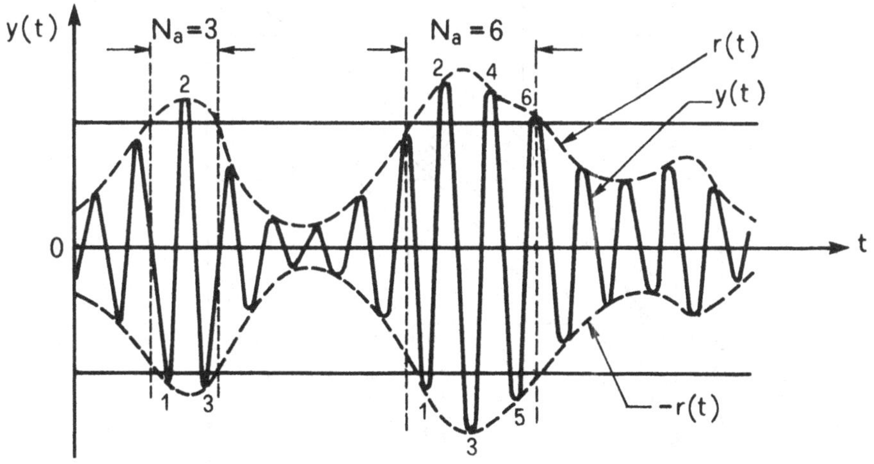

Fig. 5.1 The Envelope r(t) of the Stationary Random Process y(t)

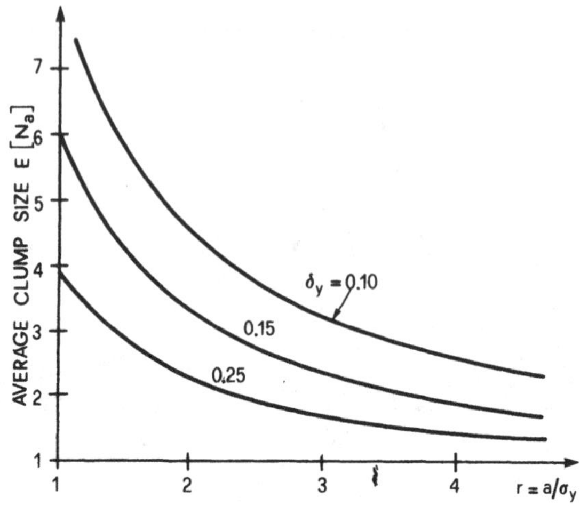

Fig. 5.2 The Average Clump Size as a Function of the Reduced Threshold Level $r = a/\sigma_y$ and the Bandwidth Factor δ_y

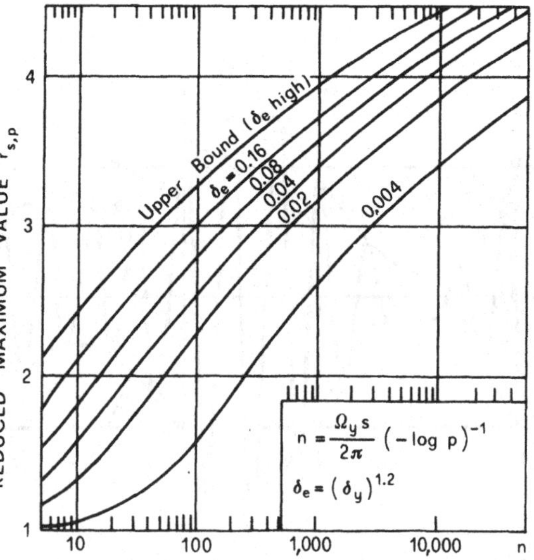

The Reduced Maximum Value $r_{s,p}$ as a Function of $n = (\Omega_y s/2\pi)(-\log p)^{-1}$ for Several Values of $\delta_e = (\delta_y)^{1.2}$

AN APPROACH TO CHARACTERIZING, MODELING AND ANALYZING EARTHQUAKE EXCITATION RECORDS

FRANK KOZIN
Polytechnic Institute of New York

1. INTRODUCTION

In these lectures we shall briefly describe an approach to the study of strong motion earthquake accelerograms that treats them basically as non-stationary time series. We are motivated by three distinct problems:

(1) The statistical problem of modeling general non-stationary time series.

(2) Characterization of earthquake acceleration records as non-stationary stochastic processes.

(3) Predict the dynamic characteristics of local ground surface behavior from strong motion records by means

of the stochastic model obtained in (2).

In these lectures we shall discuss problems (1) and (2)*).

2. BRIEF REVIEW OF EARTHQUAKE MODELS

The modeling of strong motion earthquake records as stochastic processes has been a significant topic of study at least since the early 1950's. The main objective of such studies has been to determine the basic features of earthquake excitations in order to achieve more reliable design procedures for structures that must safely withstand such natural forces. This is especially important for dams as well as nuclear power plants, whose failure or destruction during an earthquake could have catastrophic effects on the neighboring communities and environment.

The reason for the long time span devoted to these studies is due in part to the lack of sufficient numbers of strong motion records, upon which researchers could perform

*) The material presented in these lectures has been developed in collaboration with Dr.Richard Gran of the Research Division of Grumman Aerospace Corporation and with my student, Dr.T.S.Lee, under a research grant GI-43095 from the RANN Division of the National Science Foundation.

studies, prior to 1970, and in part due to the lack of robust statistical techniques through which one can study non-stationary time series.

Thus, it has been difficult to obtain a general approach to the study of these data. However, the availability of strong motion records has changed dramatically since 1970. This is basically due to the San Fernando earthquake of February 1971, of which over 200 records were obtained. These include 175 records from the Los Angeles area, 20 to 50 kilometers from the epicenter. Of these 175 records, 170 are from 57 high rise buildings which are required by a California building code to have strong motion recording instruments installed in the basement, the roof and some intermediate floor of all buildings over six floors high.

This earthquake also yielded the first near field strong motion record ever obtained, within a few kilometers of the epicenter with recorded peak accelerations exceeding 1g. The complete set of records both corrected, uncorrected as well as associated response and Fourier spectra is available in reports and on magnetic tape through the Cal Tech Earthquake Engineering Research Laboratories[1].

We have stated above that the earthquake excitation is basically a statistically non-stationary phenomenon. It is non-stationary in its amplitude statistics as well as its frequency statistics.

In Figure 1 - Figure 4, we show four typical earthquake records. It is clear from these records that the strong motion excitation cannot be separated out and identified as a statistically stationary portion of the record. Although, in lieu of other available techniques at this time, it may be expedient from an engineering point of view to treat the recording on this basis, as advocated by a number of researchers[2,3], it is our feeling that much will be gained from looking at the entire record as one non-stationary phenomenon and that is the basis upon which we have proceeded in our studies. A number of non-stationary models for the simulation of strong motion-like records have appeared over the past fifteen years.

For example, Bogdanoff et al[4] proposed a model of the form

$$\ddot{Z}_g(t) = \sum_{j=1}^{n} t\, b_j \exp(-\alpha_j t)\cos(\omega_j t + \varphi_j) \tag{2.1}$$

where $\{b_j\}$, $\{\alpha_j\}$, $\{\omega_j\}$ are given positive constants and the $\{\varphi_j\}$ is a sequence of independent identically distributed random variables, uniformly distributed on $[0,2\pi]$. This model yields sample functions which are non-stationary in amplitude as well as frequency.

A model of non-stationary stochastic excitation that has appeared a number of times in the literature[13] is simply a deterministic amplitude modulated stationary process. That is

$$\ddot{Z}_g(t) = \varphi(t)f(t), \qquad (2.2)$$

where $\varphi(t)$ is a given time function and $f(t)$ is a stationary process, such as white noise, filtered white noise, or a shot noise. Although, (2.2) may fit amplitude statistics for real strong motion earthquakes, it cannot yield the time varying frequency characteristics and hence cannot be considered as a satisfactory model, even though its simplicity is inviting.

Shinozuka and Sato[5] as well as Jennings, Housner and Tsai[6] have considered the model,

$$\ddot{Z}_g = \int_{-\infty}^{\infty} h(t - t)\varphi(\tau)W(\tau)d\tau \qquad (2.3)$$

for earthquake simulations.

They have assumed an envelope $\varphi(t)$ of the form

$$\varphi(t) = h(e^{-\alpha t} - e^{-\beta t}), \qquad (2.4)$$

and $W(t)$ is gaussian white noise. We shall see later that φ in (2.4) is not an effective envelope for fitting real earthquake records.

Cornell[7] as well as Lin[8], have proposed filtered shot noise models of the form,

$$\ddot{Z}_g(t) = \sum_{n=1}^{N(t)} G(t - t_n, A_n), \qquad (2.5)$$

where, $(N(t),\ t \geq 0)$ is the Poisson counting process, $G(t)$ is a suitable pulse with a random amplitude parameter A_n, and t_n are the occurrence times of the pulse which are Poisson distributed.

Among other models of significance related to the above approaches are those due to Amin and Ang[9], Goto and Toki[10]. Finally, Liu[3] breaks the record into three separate phases, each as stationary processes.

An interesting feature of[5] is that in the form of the model (2.3) the impulse response h(t) for the filter is chosen so that $E\{\ddot{Z}_g(t)\}$ will eventually tend to zero. This is an important physical requirement of any acceleration model. It is interesting to note that non-zero residual velocities are present in most other models cited above.

The most recent approach along these traditional lines appears in[11] and more comprehensively in[12]. In these papers which were written before the major increase in the availability of strong motion data, we postulated four criteria for determining the validity of a proposed simulation process.

(1) Non-Stationary

The zero mean simulation process should possess acceleration and velocity variance functions that ascend to a maximum and then decay to a small value with increasing time.

(2) Covariance

Assuming stationary during the period of strongest motion, the normalized covariance function of the acceleration during that period should fit envelopes and characteristics of covariances obtained from existing strong motion records.

(3) Maximum Ground Acceleration

Simulated earthquakes that are scaled to presumed intensities of former earthquakes should possess similar values for the maximum ground acceleration.

(4) Response Spectra

The Response Spectra for the simulated records, when normalized to a common spectrum intensity, should be similar to the standard spectra.

These four criteria appeared to be the main properties of significance from both a statistical as well as an engineering point of view. This is especially true in view of the fact that few strong motion earthquake records were available.

None of the models presented up to that point could satisfy these four criteria. We found that it was possible, by a simple conceptual generalization of previous models, to construct a simulation process that fit all four of the criteria.

Whereas the simulation in[5] can be looked at schematically as

$$W(t) \longrightarrow \overset{\varphi(t)}{\overset{\downarrow}{\otimes}} \longrightarrow \boxed{h} \longrightarrow \ddot{Z}_g,$$

the only change in[11,12] is to replace W(t) by a filtered noise input source N(t), yielding

$$N(t) \longrightarrow \overset{\varphi(t)}{\overset{\downarrow}{\otimes}} \longrightarrow \boxed{h} \longrightarrow \ddot{Z}_g,$$

where

$$N(t) = \int_0^C \psi(\xi)W(t - \xi)d\xi, \qquad (2.6)$$

ψ is chosen appropriately and C is a fixed upper limit.

The complete details of how well this approach fits the criteria as well as some applications can be found in[12] and[14].

It is an interesting fact that of the contributions that we have referred to so far not one comes from a statistical time series specialist. Indeed, most of the simulation procedures that we have cited have been motivated by being able to create records that look like earthquakes and satisfy one or more practical criterion. On the other hand, statisticians have been heavily involved in the analysis of seismic records. We can mention in this connection E.A.Robinson whose work has been motivated by analyzing seismic records for detection and discrimination between underground nuclear explosions and earthquakes of similar magnitudes[15], and by the problem of detecting possible sources of oil from seismic recordings[16].

In the next section we shall approach the problem of simulating and characterizing earthquakes somewhat differently than in the past. We shall apply the data directly to estimate a time varying model that appears to have promise not only for dynamical studies of structures by realistic simulations, but it also appears to have the potential to associate and study geophysical properties through its nonstationary parameters.

3. A MODEL FOR STRONG MOTION EARTHQUAKES

The general model for the earthquake process that we consider is given by the equation

$$x^{(n)}(t) + a_{n-1}(t)x^{(n-1)}(t) +$$

$$a_1(t)\dot{x}(t) + a_o(t)x(t) = \varphi(t)W(t), \qquad (3.1)$$

where x(t) representents the earthquake record, (acceleration, velocity or displacement), W(t) is the Gaussian white noise, φ(t) is an amplitude modulation on the white noise input, to be determined, and the a_j's are unknown time varying coefficients.

Although the procedures we shall outline will hold for any order differential equation (3.1), we have so far limited our studies to the second order differential equation, that is, n = 2. We are motivated to study the second order model by the ground transfer function of Kanai[2] as well as recent work due to Beaudet[17]. Hence in what follows we shall be concerned with a model of the form

$$\ddot{x}(t) + a(t)\dot{x}(t) + b(t)x(t) = \varphi(t)W(t) \tag{3.2}$$

Clearly, a(t), b(t) cannot be arbitrary unknown functions nor can the amplitude modulation $\varphi(t)$ be arbitrary.

In our procedure, we select $\varphi(t)$ initially. There is a structural reason for this that will be discussed later.

We fit a curve to the envelope of the strong motion record. We have applied specific functional forms such as

(i) $kte^{-\alpha t}$

(ii) $k(e^{-\alpha t} - e^{-\beta t})$

(iii) $\sum_{i=1}^{K} k_i \, e^{-\alpha_i t}$

on a least squares basis, but have found these to be unsatisfactory for fitting the envelopes of strong motion records. (i) gives the worst fit and (iii) gives the best depending upon how many terms are taken. The problem we have found is that in order to obtain a good fit to the initial rise in the envelope, the curve cannot match the decay rate for most records. Instead of this, we have chosen to generate an empirical envelope which is fitted by a spline technique. Thus, cubic polynomials are fit to the maxima of the envelope in such a way that the derivatives are matched at the end points where different polynomials meet as illustrated in Figure 5.

In Figure 6 and Figure 7 we present two examples of envelopes obtained by spline fitting to the strong motion records of Figure 3 and Figure 4. We note a rather sharp change in the envelope at the point at which the strong motion portion initiates and terminates. It is conceivable that this type of envelope study may yield a more accurate definition of the duration of the strong motion portion of the record.

The envelopes obtained by spline fitting techniques follow the actual physical record much more closely than by fitting the exponentials (i) - (iii).

Hence, in our model (3.2) we have obtained $\varphi(t)$ empirically and we must now consider the coefficients $a(t)$, $b(t)$. In our numerical studies we have set $a(t) \equiv a_o$, constant damping, and $b(t) = b_3 t^3 + b_2 t^2 + b_1 t + b_o$, a cubic polynomial, where a_o, b_3, b_2, b_1, b_o are unknown constants to be estimated by some procedure directly from the strong motion data.

In order to make the estimation problem more realistic, we shall also assume that the observed, that is, the given strong motion data is noisy. Hence, if x represents the strong motion acceleration, then the given data, denoted by y(t) is assumed to be of the form

$$y(t) = x(t) + \sigma v(t), \tag{3.3}$$

where v(t) represents a gaussian white noise that is independent of the white noise W(t) in (3.2).

Upon defining the vector

$$z(t) = \begin{bmatrix} x(t) \\ \dot{x}(t) \\ a_o \\ b(t) \\ \dot{b}(t) \\ \ddot{b}(t) \\ \dddot{b}(t) \end{bmatrix} = \begin{bmatrix} z_1 \\ z_2 \\ z_3 \\ z_4 \\ z_5 \\ z_6 \\ z_7 \end{bmatrix} \tag{3.4}$$

where $b(t) = b_3 t^3 + b_2 t^2 + b_1 t + b_o$, we can pose our problem as follows. Estimate z(t) for the system

$$\begin{aligned} &(a) . \quad \frac{dz(t)}{dt} = F(z)z(t) + b(t)W(t), \\ &(b) . \quad y(t) = Mz(t) + \sigma v(t), \end{aligned} \tag{3.5}$$

where σ is a scalar,

$$F(z) = \begin{bmatrix} 0 & 1 & 0 & 0 & 0 & 0 & 0 \\ -z_4 & -z_3 & 0 & 0 & 0 & 0 & 0 \\ 0 & 0 & 0 & 0 & 0 & 0 & 0 \\ 0 & 0 & 0 & 0 & 1 & 0 & 0 \\ 0 & 0 & 0 & 0 & 0 & 1 & 0 \\ 0 & 0 & 0 & 0 & 0 & 0 & 1 \\ 0 & 0 & 0 & 0 & 0 & 0 & 0 \end{bmatrix}, \quad b(t) = \begin{bmatrix} 0 \\ \varphi(t) \\ 0 \\ 0 \\ 0 \\ 0 \\ 0 \end{bmatrix}$$

and $M = [1,0,0...0]$, given the observations $\{y(s);\ 0 \leq s \leq t\}$.

We denote the estimator of $z(t)$ by $\hat{z}(t)$. Furthermore, we shall choose $\hat{z}$ to be the least squares estimator. That is, $\mathfrak{z}(t) = \hat{z}(t)$, minimizes the expectation

$$E\left\{ \| z(t) - \mathfrak{z}(t) \|^2 \right\}.$$

The theoretical solution for $\hat{z}(t)$ as a functional of the observed data $\{y(s);\ 0 \leq s \leq t\}$ is known to be the conditional expectation,

$$\begin{aligned} \hat{z}(t) &= E\{z(t) \mid y(s);\quad 0 \leq s \leq t\} \\ &= \int z\ p(z,t \mid y(s); 0 \leq s \leq t)\,dz. \end{aligned} \tag{3.6}$$

In the general case it is not possible to proceed any further that the formal solution (3.6). However, for the system (3.5) that we are considering one can proceed.

Indeed, if in (3.5)(a) we formally set $W(t) = \frac{dB(t)}{dt}$, where B(t) is the Brownian Motion, and multiply the equation by dt, we obtain the Ito differential equation[18]

$$dz(t) = F(z(t))z(t)dt + b(t)dB(t). \qquad (3.7)$$

The solution process of (3.7) is a diffusion process and its conditional density function $p(z,t|\xi,\tau)$, for $\tau < t$, is the solution of the Fokker-Planck, or forward diffusion equation,

$$\frac{\partial p}{\partial t} = \mathcal{L}^* p \qquad (3.8)$$

$$\equiv - \sum_{i=1}^{N} \frac{\partial}{\partial z_i}\left[(F(z)z)_i p\right] + \frac{1}{2} \sum_{i,j=1}^{N} \frac{\partial}{\partial z_i \partial z_j}\left[(b(t)b^T(t))_{ij} p\right]$$

where $(\ )_i$ denotes the ith component of the vector $F(z)z$, and $(\ )_{ij}$ denotes the (i,j) element of the matrix $b(t)b^T(t)$. We have ssumed $E\{B^2(t)\} = t$, and $N = 7$.

The equality (3.6) shows us that we must obtain the conditional density, $p(z,t|y(s);\ 0 \leq s \leq t)$.

It is known[19] that the conditional density for the system (3.5) satisfies the stochastic partial differential equation

$$dp = \mathcal{L}^* p\ dt + \frac{1}{\sigma^2}\left[(z - \hat{z})^T M^T (dy - M\hat{z}\ dt)\right] p, \qquad (3.9)$$

where M, σ are defined in (3.5), $\hat{z}$ is defined by (3.6), and $\mathcal{L}^*$ ist the forward diffusion operator defined in (3.8).

The adjoint of the forward diffusion operator $\mathcal{L}^*$ is denoted by $\mathcal{L}$. The operator $\mathcal{L}$ is the backward diffusion operator, which is often referred to as the generator of the Markov process.

For any function h(z), which has a finite expectation, we have the identity,

$$\int h(z)\left[\mathcal{L}^* p(z,t\,|\,y(s);\ 0 \leq s \leq t)\right] dz$$

$$= \int\left[\mathcal{L}\, h(z)\right] p(z,t\,|\,y(s);\ 0 \leq s \leq t) dz \qquad (3.10)$$

$$= \widehat{\mathcal{L}h(z)} \ ,$$

where "$\widehat{\ }$" denotes expectation conditional upon the observation process $\{y(s);\ 0 \leq s \leq t\}$. Hence, upon multiplying (3.9) by h(z), integrating and then applying the notation of (3.10) we obtain,

$$d\hat{h}(z) = \widehat{\mathcal{L}\, h(z)} dt$$

$$+ \frac{1}{6^2}\left(\widehat{zh(z)} - \hat{z}\widehat{h(z)}\right)^T M^T (dy - M\hat{z}\ dt) \qquad (3.11)$$

which is an equation for the conditional expectation,

$$E\{h(z(t))\,|\,y(s);\ \ 0 \leq s \leq t\}.$$

We can now obtain equations for the conditional moments,

$$\hat{z}_i,\ \ \widehat{z_i - \hat{z}_i},\ \ \widehat{(z_i - \hat{z}_i)(z_j - \hat{z}_j)}\ ,$$

etc., where the subscripts denote the components of the

z-vector.

Since (3.11) is a non-linear differential equation, the equations for the conditional moments will be coupled to the next higher moment.

Thus, if we denote the conditional covariance matrix by $\hat{\theta}$, the matrix of second moments,

$$\hat{\theta} = \left(\widehat{(z_i - \hat{z}_i)(z_j - \hat{z}_j)}\right),$$

then the equation (3.11) for the elements of $\hat{\theta}$ will contain the third conditional moments.

At this point, we make an approximation to the equation (3.11). The approximation is due to Schwartz and Bass[20]. Upon denoting the vector $F(z)z$ by $f(z)$, we expand the ith component, $f_i(z)$ up to second order terms about $\hat{z}$, yielding

$$f_i(z) \approx f_i(\hat{z}) + \sum_{j=1}^{N} f_{ij}^{(1)}(\hat{z})(z_j - \hat{z}_j) \qquad (3.12)$$

$$+ \sum_{j,k=1}^{N} f_{ijk}^{(2)}(\hat{z})(z_j - \hat{z}_j)(z_k - \hat{z}_k),$$

where

$$f_{ij}^{(1)} = \frac{\partial f_i}{\partial z_j}, \qquad f_{ijk}^{(2)} = \frac{\partial^2 f_i}{\partial z_j \partial z_k}.$$

Taking the conditional expectation of the equation (3.12) gives

$$\hat{f}_i(z) = f_i(\hat{z}) + \frac{1}{2} \sum_{j,k=1}^{N} f_{ijk}^{(2)} (\hat{z}) \widehat{\left[(z_j - \hat{z}_j)(z_k - \hat{z}_k)\right]} \tag{3.13}$$

$$= f_i(\hat{z}) + \frac{1}{2} \sum_{j,k=1}^{N} f_{ijk}^{(2)} (\hat{z}) \hat{\theta}_{jk} ,$$

where θ_{jk} ist the j,k element of the conditional covariance matrix.

Using (3.11) we obtain the explicit equations for the first and second conditional moments as,

$$d\hat{z}_i = \widehat{(F(z)z)_i} dt + \frac{1}{\sigma^2} \hat{\theta}_i M^T(dy - M\hat{z}dt) \tag{3.14}$$

and,

$$d\hat{\theta}_{ij} = 2 \widehat{\left[(F(z)z)_i z_j - \widehat{(F(z)z)_i}\hat{z}_j\right]}_{sym} dt \tag{3.15}$$

$$+ \frac{1}{\sigma^2}\left[\hat{\theta} M^T M\hat{\theta}^T + bb^T\right]_{ij} dt, \text{ respectively,}$$

where $(F(z)z)_i$ is the ith component of the vector $F(z)z$, θ_i is the ith row vector of the matrix θ, θ_{ij} is the (i,j) element of the matrix θ and $[\ ]_{sym}$ denotes the process of symmetrically carrying out the indicated products. Finally, since $F(z)z$ is quadratic in the components of z, it follows that no third order moments will be present in (3.15). Hence, (3.14), (3.15) constitute a closed set of equations that can be solved recursively for the desired estimates.

We stated in Section III that the amplitude modulation

function $\varphi(t)$ must be selected initially and that there is a structural reason behind this fact. The reason is that the filter equations (3.14), (3.15) derived from the Schwarz-Bass approximation technique do not allow estimates for the function $\varphi(t)$. Indeed, $\varphi(t)$ is not an observable variable for this approximate filter. Hence, it must be chosen according to some other criterion, as described in Section III.

One illustration of how well this approximate filter works for a constant coefficient example is given in figure 8.

We simulated the system,

$$\begin{aligned} \ddddot{x} + 22.5\dddot{x} + 20\ddot{x} + 16\dot{x} + 8x &= W \\ y &= x + 0{,}1v, \end{aligned} \tag{3.16}$$

where W, v are independent white noise terms with unit intensities. Estimating the coefficient of $\ddot{x}$ by the equations (3.14), (3.15) gave us results as illustrated in figure 8. In this case the true coefficient is -20. We see that the first pass using zero initial conditions rapidly converges to a small fluctuation about the true value. Using new initial conditions the fluctuation about -20 becomes smaller.

There are no convergence theorems available upon which to judge the accuracy of the estimations obtained through the approximate equations (3.14), (3.15).

On the other hand it is known that for the true optimal estimator z, the so-called innovations process $dy - M\hat{z}\,dt$, should be a white noise[21]. Thus, we can determine the quality of the estimated parameters by applying statistical tests to the innovations process.

We apply tests such as the run test and the trend test to determine the whiteness of the innovations. We accept an estimate solely upon whether the result of each test lies within specified bounds or confidence limits[22]. We also estimate the spectral density of the residuals as a further test for whiteness.

In figures 10 and 11, sample functions of the earthquake model (3.2) generated by the data from the actual records of figures 3 and 4 are presented. It can be seen that they contain a large portion of the significant nonstationary characteristics of the original records.

Our research program is continuing along the lines we have described in these lectures in order to obtain a reliable modeling procedure with the ultimate goal of predicting ground motion characteristics for microzonation applications.

REFERENCES

[1] Strong motion earthquake accelerograms, Tech. Rep. Earthquake Engineering Research Lab., Cal. Inst. of Tech., Vol.I, 1970, Vol.II, 1973

[2] Vanmarcke, E.H., Structural response to earthquakes, Chapter 8 in Seismic Risk and Engineering Decisions, C. Lomnitz and E. Rosenblueth, Eds., Elsevier Book Co., Amsterdam, 1976.

[3] Liu, S.C., Synthesis of stochastic representations of ground motions, Bell System Technical Journal, 521, April 1970.

[4] Bogdanoff, J., et al., Response of a simple structure to a random earthquake type disturbance, Bull. Seis. Soc. Amer., 51, 293, 1961.

[5] Shinozuka, J. and Sato, Y., Simulation of nonstationary random processes, Proc. ASCE, 93, EM1, 11,1967.

[6] Jennings, P., et al., Simulated earthquake motions. Tech. Rep. Earth. Eng. Res. Lab., Cal.Tech. April 1968.

[7] Cornell, C., Stochastic process models in structural engineering. Dept. Civil Eng., Tech. Report No.34, Stanford, 1964.

[8] Lin, Y.K., On nonstationary shot noise, Jour. Acous. Soc. Amer., 36, 82, 1964.

[9] Amin, M. and Ang, A., Nonstationary stochastic model of earthquake motions, Proc, ASCE, 94, EM2, 559, 1968.

[10] Goto, H. and Toki, K., Structural response to nonstationary random excitation, Proc. 4th WCEE, Santiago Chile, 1, A-1, 130, 1969.

[11] Levy, R. and Kozin, F., Processes for earthquake simulation, Proc.ASCE, Journ. Eng. Mech. 94, EM6, 1597, Dec. 1968.

[12] Levy, R. et al., Random processes for earthquake simulation, Proc. ASCE, 97, EM2, 495, 1971.

[13] Bolotin, V.V., Statistical Theory of the aseismic design of structures, Proc. 2nd WCEE, Tokyo, Japan, 1365, 1960.

[14] Grossmayer, R., On the application of various crossing probabilities in the structural aseismic reliability problem, Proc. IUTAM Symp. Stoch. Prob. in Dynamics, Southampton, 17, July 1976.

[15] Robinson, E.A., Mathematical devolopment of discrete filters for the detection of nuclear explosions, Jour. Geophys. Res., 68, 19, 5559, 1963.

[16] Robinson, E.A., Predictive decomposition of time series with applications to seismic exploration, Geophysics, 32, 418, 1967.

[17] Kozin, F. and Gran, R., An approach to analysis and modeling of earthquake data in Stochastic Problems in Mechanics, Univ. of Waterloo Press, 193, 1974.

[18] Gran, R. and Kozin, F., Nonlinear filtering applied to the modeling of aerthquake data, Proceedings of the Symposium on Nonlinear Estimation and Its Applications, San Diego, Calif., September 1973.

[19] Bass, R. and Schwartz, L., Optimal multichannel non-linear filtering, J.Math. Anal. and Appl., 1966.

[20] Kailath, T., An innovations approach to least squares estimation, Part I, Trans. Auto. Contr., AC-13, 646, Dec. 1968.

[21] Bendat, J.S. and Piersol, A.G., Random data; Analysis and Measurement Techniques, John Wiley and Sons, New York, 1971.

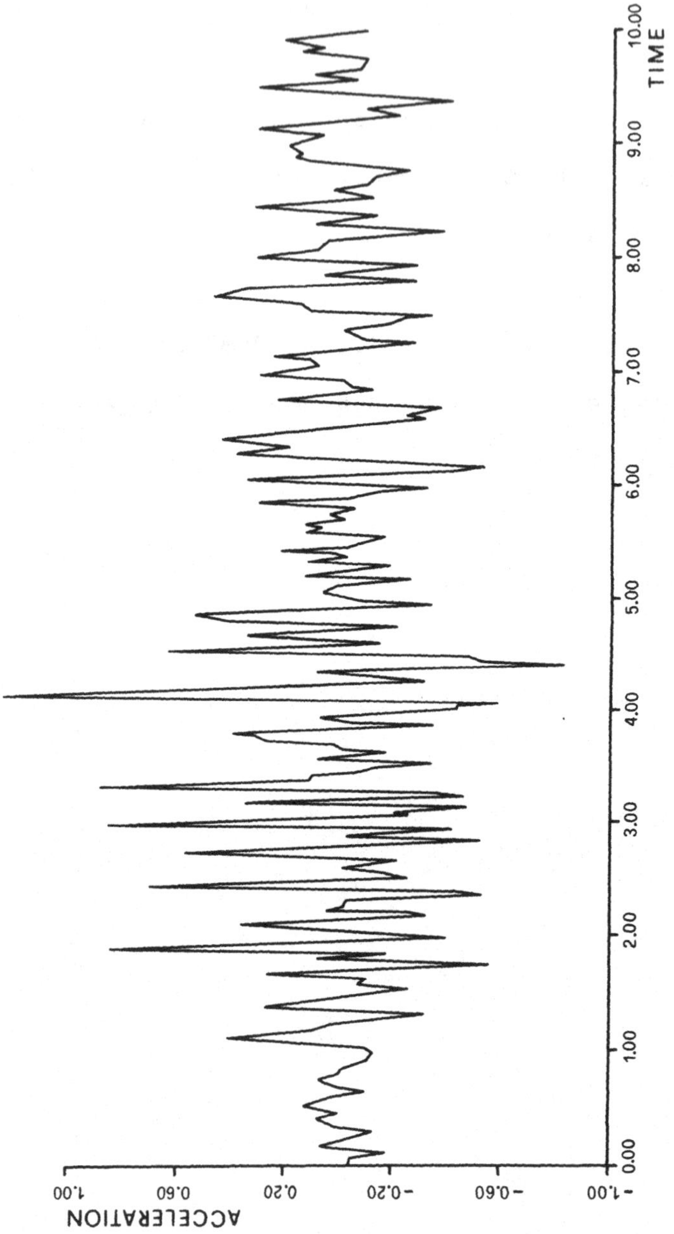

Fig. 1 Strong Motion Record

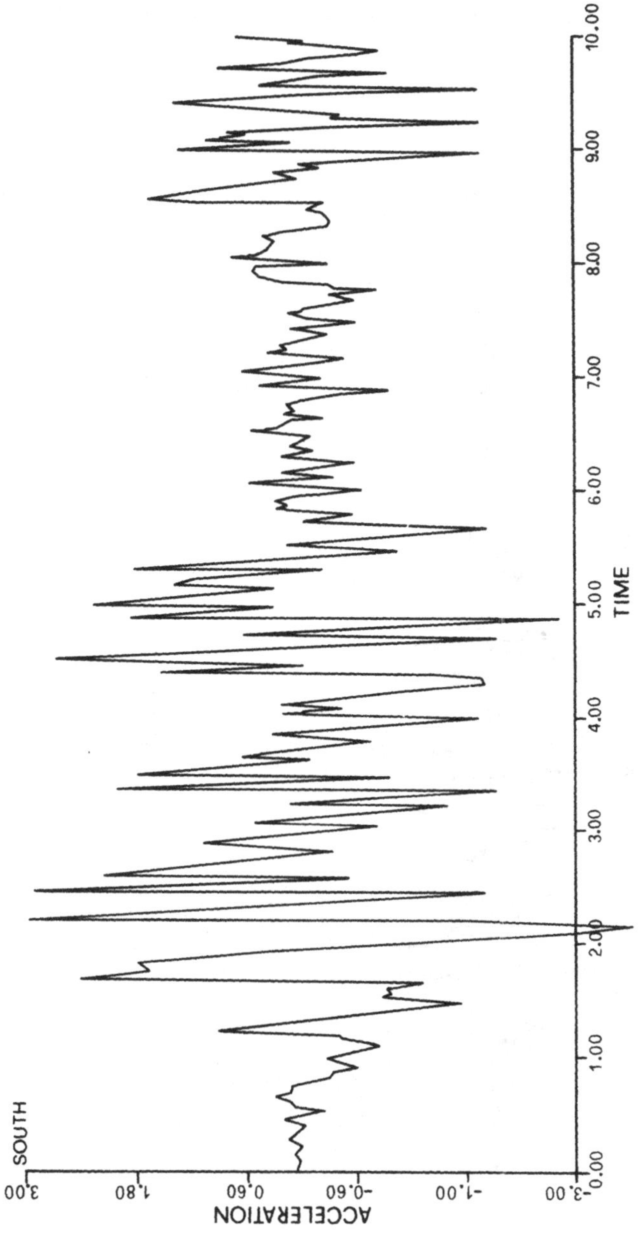

Fig. 2 Strong Motion Record

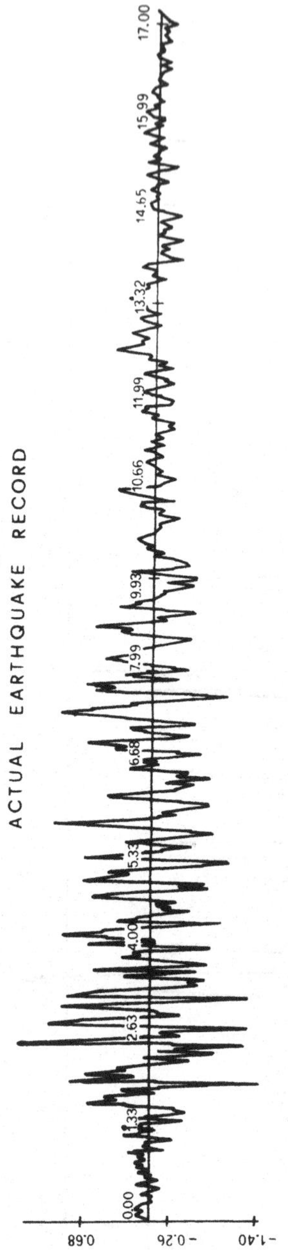

Fig. 3 Actual Earthquake Record

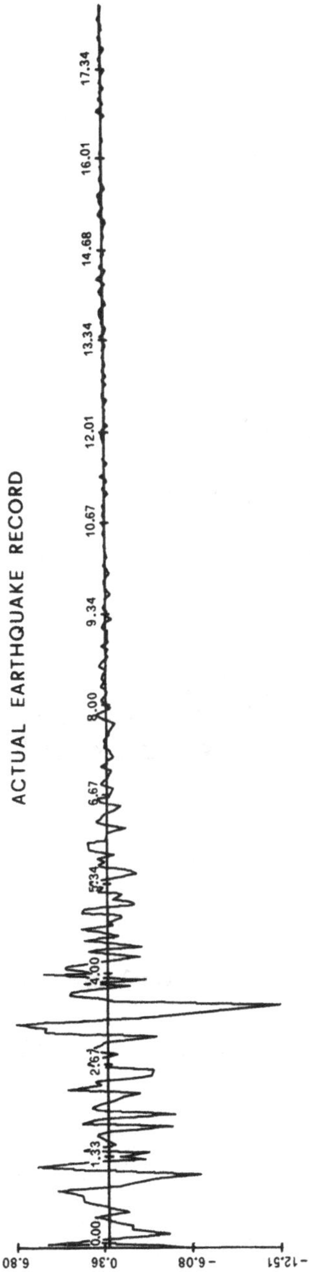

Fig. 4 Actual Earthquake Record

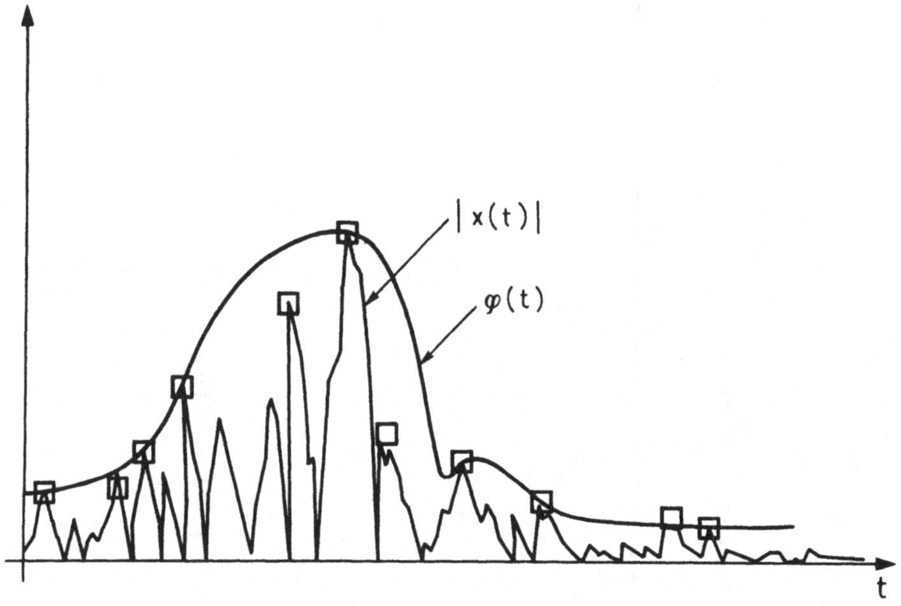

Fig. 5 Empirical Envelope

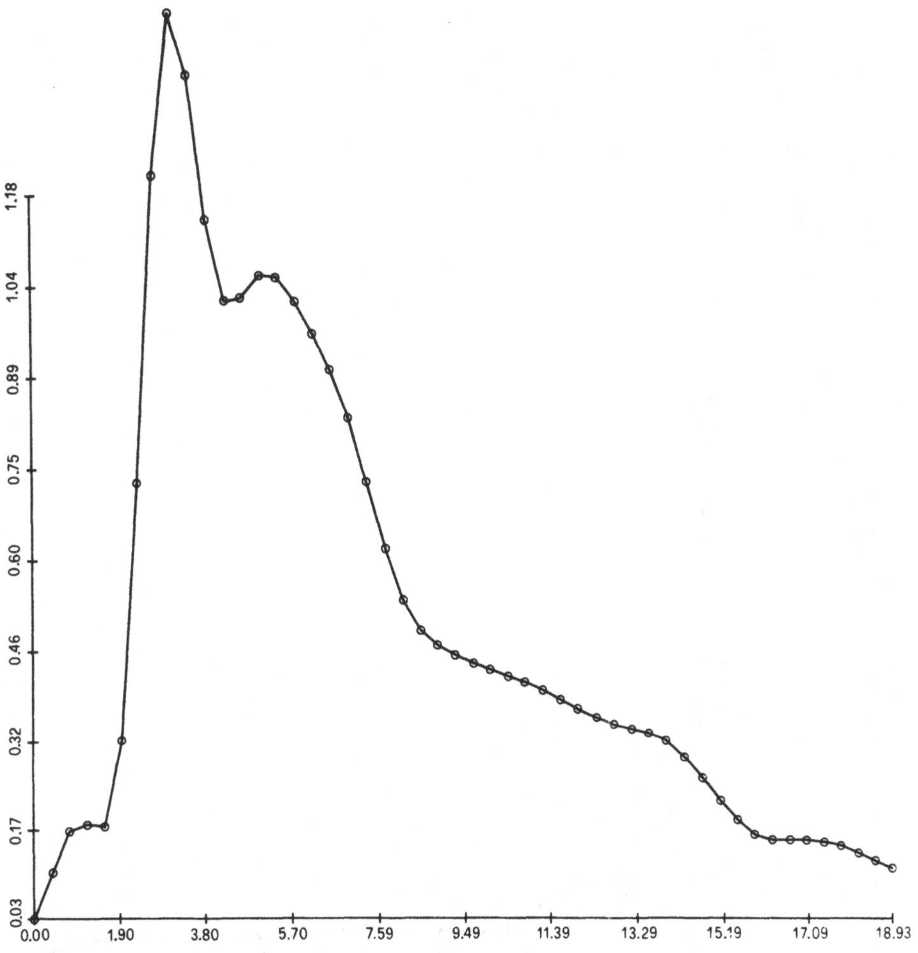

Fig. 6 Envelope obtained by Spline Fitting

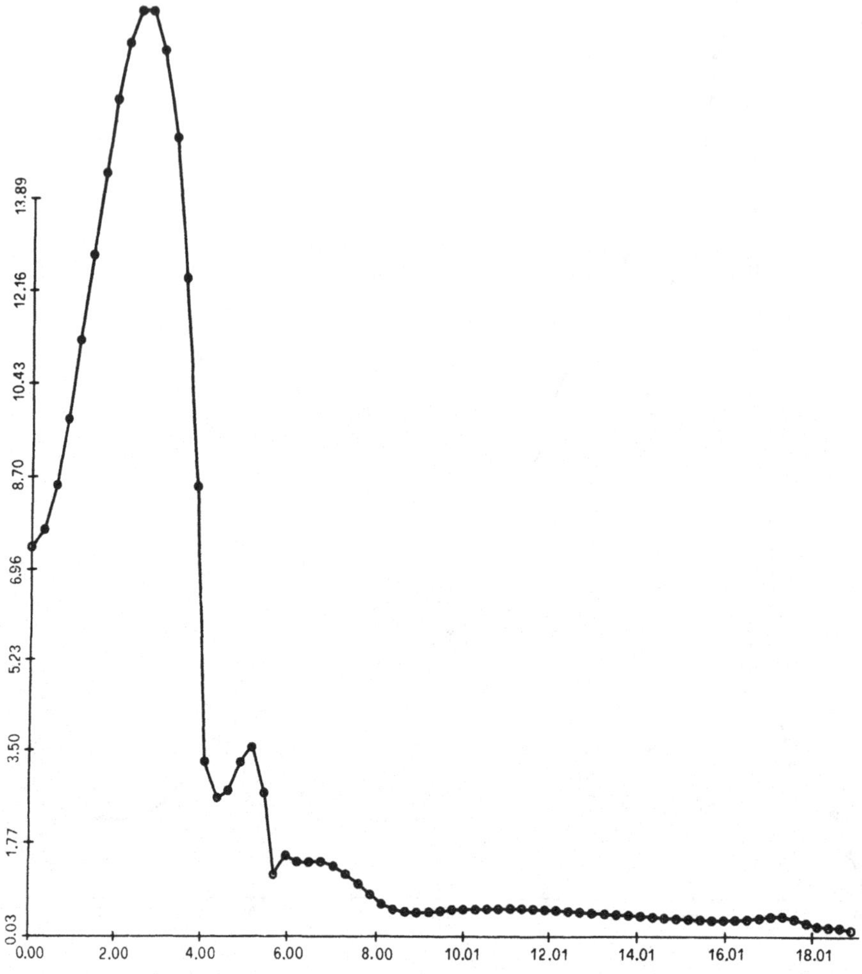

Fig. 7 Envelope obtained by Spline Fitting

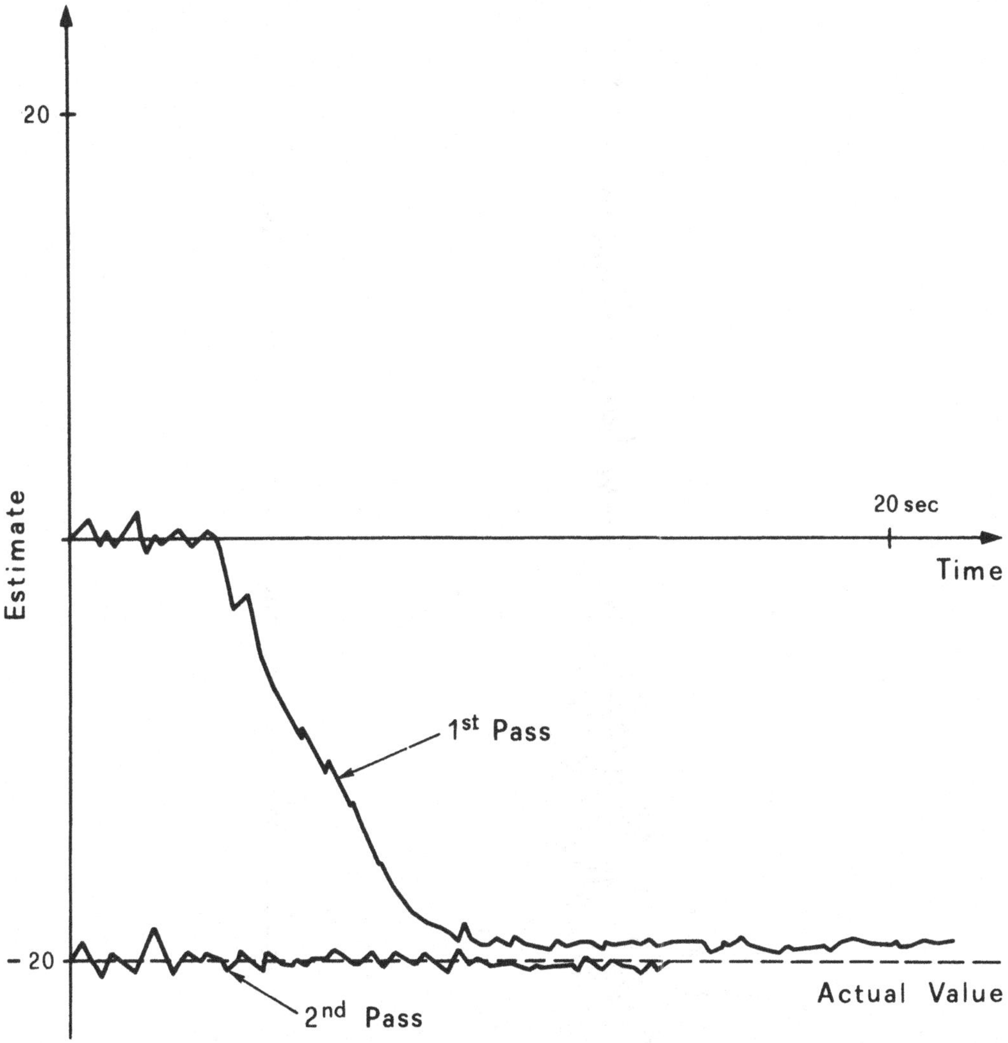

Fig. 8 Estimate of Coefficient

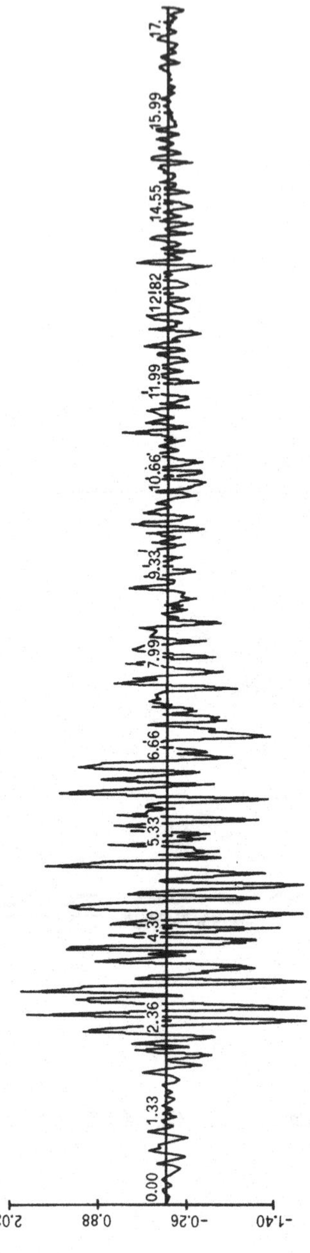

Fig. 9 Sample Function Generated by the Data of Figure 3

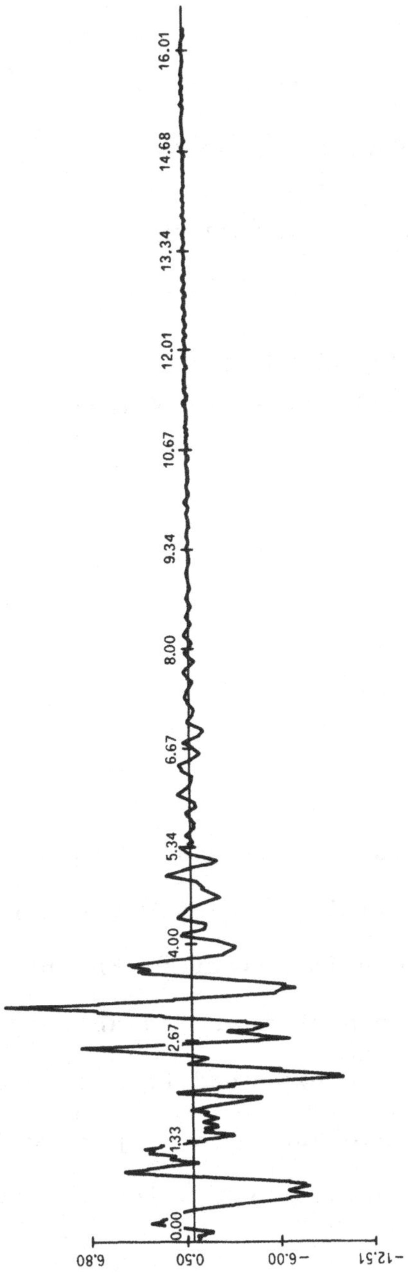

Fig.10 Sample Function Generated by the Data of Figure 4

ASEISMIC RELIABILITY
AND
FIRST-PASSAGE FAILURE

RUDOLF GROSSMAYER
Technical University of Vienna

1. INTRODUCTION

The problem of aseismic reliability is one among various problems within the field of earthquake engineering, but one of greatest inportance in practice. Engineers are expected to design structures that are capable to withstand any earthquake, that may be expected to happen at a given site. This is often an unsolvable task, particularly so, if certain restrictions like cost limitations, have to be kept in mind.

Therefore, an important question for a given design is

that of the safety of the structure against an "expected" earthquake occuring within the lifetime of the structure. In fact, this question cannot be answered absolutely, but only under certain conditions. Why?

First of all earthquakes are typically random in nature. No two earthquakes are alike. The mechanism of origin is not unique. The different types of earthquake waves are often reflected and refracted on their way from the epicenter to the site. Therefore, the excitation process seems like an irregular motion. Predictions about future quakes are based on past observations. Unfortunately, there are some difficulties arising from insufficient instrumentation. The last quake, for example, that happened in this region only a few weeks ago, was so strong, that some of the meteorological stations in middle europe did not produce a usable record, because their accelerometers were only designed for weaker quakes. This lack is not present in regions like California or Japan with a high seismic activity.

One result of past measurements is a relationship between an expected quake magnitude and occurence probability, given, for instance, in expected number of quakes arriving within thousand years. Of course, this probability decreases with increasing magnitude. The use of such relations provides a possibility for the authorities, for instance, to select a certain quake magnitude for the design specifications of the structure under consideration.

The random nature of the quake requires a random model for the mathematical description of the acceleration process. This model must be influenced by the soil properties of the site and must include all past information. The question how to find proper models is the topic of a parallel lecture and will not be discussed here.

The response of the structure under the random excitation will be random, too. Therefore, the answer to the question of safety holds with a certain probability only. This is an unusual situation for engineer and authority; but all other interpretations in a purely deterministic sense are not taking into account the real situation. Fortunately, we are not often confronted with such problems, where the random nature of the loads is so significant (mind, that nearly all phenomena in nature are random, but most of them show only small fluctuations about a well known mean and can, therefore, be treated in a deterministic sense). Hence, one of the aims of this lecture and the whole session will be to get you more familiar with such problems and to show you the different way of thinking you need for their treatment.

1.1 Classification of structures and earthquakes

Various types of structures have to withstand earthquakes. In regions where strong motion earthquakes are rare events, only special structures are designed under this point of view, e.g., power plant systems, broadcasting transmitters and sometimes tall buildings. In may cases the structure is a very complicated one and can be divided into the main structure, the so-called primary structure, and substructures, the so-called secondary structures.

With nuclear power plant systems two different types of "design earthquakes" are usually considered. First, an operating basis earthquake (OBE) is considered. It is a weak earthquake, that may almost surely be expected to happen and must not endanger the operation of the power plant. Second, a safe shutdown earthquake (SSE) is considered. This is the strongest motion expected to occur within the presumed lifetime of the plant. It is not expected, that the whole plant including all main and auxiliary equipment will survive the quake without damage. But it must be demanded, that a shutdown of the nuclear system is guaranteed without any danger for the surroundings.

Comparing these two design conditions the second is, in general, the more important and restrictive. We will therefore, concern ourselves in the following with the reliability under strong motion earthquakes.

Before we can define the reliability of a structural system under earthquake loadings we have to investigate the modes of failure, that may arise.

1.2 Modes of failure

Common to the various methods of design of structures under static loads is the assumption, that a specific value of the maximum stress (given the material properties and some safety coefficients reflecting different uncertainties in design procedure) must not be exceeded within the structure. Sometimes we have to ensure that certain deformations remain within a prescribed tolerable domain in order not to endanger the operation of the system. In many cases we have to bear in mind both conditions, and the engineer has to check which of them is the decisive one in the special case.

If we design structures under dynamic loads the limitations are the same as in the case mentioned before, but the method of calculation is more extensive. We have to investigate the dynamical properties of the structure and to study the structural response to the given excitation. The maximum deflections and stresses are sometimes very sensitive to small variations in the dynamical properties and, therefore, have to be evaluated carefully.

Another point of view arising in this connection is the question of <u>fatigue failure</u>. The structure is assumed to collapse under a specific stress level if a certain number of cycles has been reached. The reason for this mode of failure lies in material deterioration.

Applying these consideration to the aseismic reliability problem it can be concluded that

1) the question of (mainly low cycle) fatigue is interesting in the case of design under OBE. In spite of a relatively small number of quakes successive deterioration of the masonry or concrete may cause failure. But because of lack of information the actual importance of this feature has not yet been clarified,

2) it is of primary interest to consider the structure under strong motion earthquakes. In this case the first criteria mentioned above is that of practical importance. In what follows, we will restrict ourselves to this case and give the following definition.

1.3 Definition of Reliability

A structure will be called reliable, if a characteristic value X of the structural response remains with a certain probability within a prescribed tolerable domain Γ during the lifetime T_1 of the structure.

It is assumed therefore that structural collapse will occur, if the characteristic value X, being a random process in time, will exceed (the or) one of the tolerable bounds for the first time.

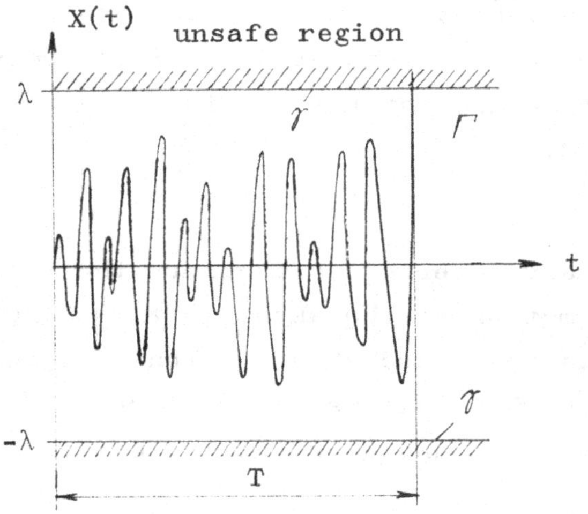

Using the notation of probability theory this event is called "first passage". The problem of aseismic reliability is therefore tantamount to the first-passage problem in probability theory. (Sometimes you find in the literature the equivalent expression "first-excursion problem").

1.4 Classification of first-passage problems

This problem, stated above in general form can be posed under various kinds of initial conditions and safety regions Γ. Without loss of generality we can assume the system to be initially at rest in many cases. We are then speaking about "zero initial conditions" in contrast to "random initial conditions", where $X(t_o)$ is a random variable. In the latter case it may happen, that the process $X(t)$ is outside of Γ at $t = t_o$. This effect is accounted for by a non-vanishing initial failure probability.

In applications, two different types of domains Γ are usually met.

1) the one-sided boundary problem

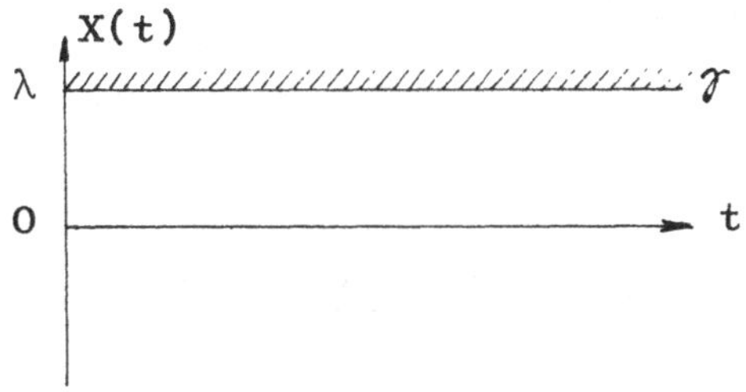

$$\Gamma_o = \{x/x \le \lambda\}$$

2) the double-sided boundary problem (for a symmetric process with zero mean).

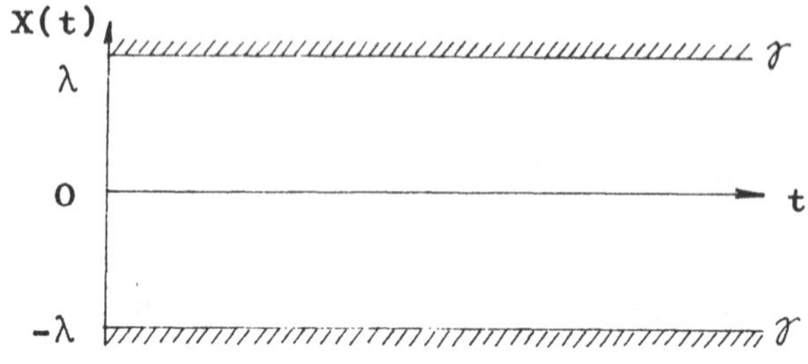

$$\Gamma_d = \{x / |x| \leq \lambda\}$$

Sometimes, also an envelope boundary problem is considered to give useful approximations for the other two cases.

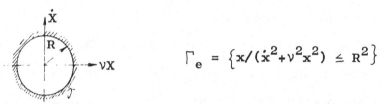

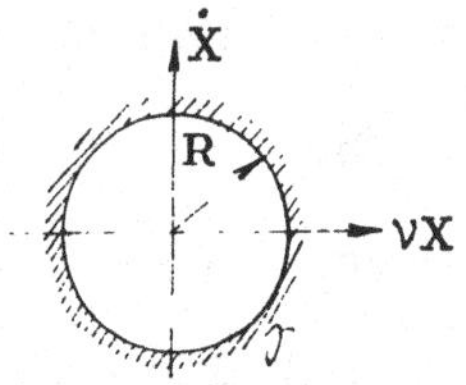

$$\Gamma_e = \{x / (\dot{x}^2 + \nu^2 x^2) \leq R^2\}$$

The first passage probability for the same process and initial conditions will always be greater for Γ_e than for Γ_d and greater for Γ_d than for Γ_o. This is intuitively clear from the relation

$$\Gamma_e \subset \Gamma_d \subset \Gamma_o$$

1.5 Basic assumptions for the response process X(t)

Before we get into more details let us make some assumptions about the excitation process $\ddot{Z}(t)$ and the response process X(t).

Generally speaking, the structural response $X(\mathfrak{r},t)$ is a function of both space and time. In order to estimate the reliability we have to find that specific point $\mathfrak{r}_o$ in space, that gives the worst result. The characteristic value of the structural response is defined by

$$X(t) = \bar{X}(\mathfrak{r}_o,t) = \max_{\mathfrak{r}} \bar{X}(\mathfrak{r},t) \qquad (1)$$

The second part of this relation holds only, if the boundary γ of the safety domain Γ is independent of space e.g. the maximum strength in the structure. The problem how to find the critical point $\mathfrak{r}_o$ is not discussed herein, as it is well known from deterministic calculations.

In contrast to wind excitation the earthquake acceleration process $\ddot{Z}(t)$ fluctuates about a zero mean. It is observed from many records, that the excitation $\ddot{Z}(t)$ is a typical nonstationary process with three different time-intervals: the starting period, the (approximately stationary) period of strong motion and the dying-out period where the excitation attenuates toward zero.

Fig.1 gives an example for a measured ground acceleration of the recent destructive earthquake in Friuli.

A widely used assumption about the acceleration process $\ddot{Z}(t)$ is that of a Gaussian probability distribution. It can be argued that according to the limit theorem for great numbers the multiple reflections and refractions of the seismic waves lead to a normal distribution. The same result is obtained in the limit for a Poisson process with a mean arrival rate ν for the independent pulses, when ν goes to infinity[1].

1.6 Structural properties and modal analysis

Statistical properties of the structural response X(t) are determined by those of the excitation $\ddot{Z}(t)$ and the dynamical properties of the structure. First of all a proper model for the structure has to be selected. In most of the cases a lumped-parameter model will do it best, e.g.Fig.2[2]. Sometimes, distributed-parameter systems are preferred or combined with lumped-parameter models. In all of these cases a linear behaviour of the structure is assumed in order to apply modal analysis. The assumptions for it are fulfilled[3], if certain linear relationships hold between the mass matrix M, the damping matrix K and the stiffness matrix C e.g. if K is proportional to C or to M^{-1}.

The structural response $\bar{X}(t)$ has to be determined from

$$M \ddot{\underset{\sim}{x}} + K \dot{\underset{\sim}{x}} + C \underset{\sim}{x} = M \ddot{z} \underset{\sim}{1} \tag{2}$$

and the excitation $\ddot{z}$ has to be measured in the direction opposite to x.

The application of modal analysis leads from equ.(2) to a system of uncoupled equations. Usually it is sufficient to consider the first modes only. Sometimes it is reasonable to take into account only the first one, thus significantly simplifying the analysis. It must be noted, that a transformation to modal coordinates is favourable in those cases, too, where the equations do not uncouple.

Structures under earthquake excitations are usually damped in various kinds. First of all internal damping can be observed. It is usually equivalent to a viscous damping with 0,5 to 1 % of critical damping. Secondly friction between parts of the primary structure and secondary structure has a damping influence on the structural response. Sometimes certain parts of the structure show hysteretic behaviour, which is taken into account either by enlarging the damping ratio and linear analysis or by a nonlinear analysis.

Summarizing all phenomena a fraction of critical damping is usually found within the range of .5 to 7 % and viscous damping is assumed. It can be concluded, that modal analysis can be applied, therefore, at least approximately by neglecting the small coupling expressions[3].

Hence, a solution of equ.(2) can be obtained by summarizing over all modal responses

$$\underset{\sim}{x} = \sum_{j=1}^{n} b_j f_j(\vartheta_o) x_j(t) = \sum_{j=1}^{n} \alpha_j x_j(t) \tag{3}$$

where n is the number of degrees of freedom, b_j is the mode participation coefficient, $f_j(\vartheta_o)$ are the eigenfunctions corresponding to the modal response $x_j(t)$, measured at $\vartheta = \vartheta_o$ according to relation (1).

The modal responses are solutions of the equations

$$\ddot{x}_j + 2\mu_j \dot{x}_j + \nu_j^2 x_j = \ddot{z}(t) \tag{4}$$

where ν_j is the undamped and $\nu_{jd} = \nu_j \sqrt{1 - \xi_j^2}$ the damped natural frequency. ξ_j in the fraction of critical damping and $\mu_j = \xi_j \nu_j$. Using the Duhamel integral the solution can be obtained from

$$x_j(t) = \int_o^t h_j(t-\tau)\ddot{z}(\tau)d\tau \tag{5}$$

where the impulse response function belonging to (1)

$$h_j(t) = \frac{e^{-\mu_j t}}{\nu_{jd}} \sin \nu_{jd} t \tag{6}$$

has to be inserted.

It follows, that for a stationary or nonstationary

Gaussian process $\ddot{Z}(t)$ with zero mean the structural response process X(t) has the following properties:

Gaussian probability distribution

zero mean

nonstationary (for stationary input a stationary steady-state solution is found after the transient response has died out)

and it will be assumed that the correlation function is twice differentiable (this holds, e.g., for the response to white noise of a linear damped oscillator).

Before going into further details about the first-passage probability, specific models have to be chosen for $\ddot{Z}(t)$. The statistics of the structural response corresponding to the excitation processes

SW - stationary white noise

NW - nonstationary white noise

FNW - filtered nonstationary white noise

are derived in the appendix.

2. EXACT OR NUMERICAL SOLUTIONS OF THE FIRST-PASSAGE PROBLEM

Up to now we have defined and classified the problem. In the following different ways of solutions will be presented.

We are looking for the first-passage probability $H(t;\lambda)$ of the structural response process $X(t)$ defined above. Because of its great importance this problem has attracted many workers within the last years. Solutions for the case of a stationary process have been found first. Those for nonstationary processes are not older than about ten years.

2.1 Exact solutions of the Pontryagin's equation

First of all we are trying to find exact solutions for each problem. In this case one possible way is to take into account the Markov property of the vector process

$$\begin{pmatrix} x(t) \\ \dot{x}(t) \end{pmatrix}$$

The process is Markov if the excitation process $\ddot{Z}(t)$ is broadband and modeled from white noise. Then, a partial differential equation can be set up for the reliability function $U(t;\lambda, \underset{\sim}{x})$, where $\underset{\sim}{x}$ is the starting position vector of the process $X(t)$ at $t = 0$ in the phase space. This equation[4,5]

$$\frac{\partial U}{\partial t} = - \alpha_i(t) \frac{\partial U}{\partial x_i} - \frac{1}{2} \beta_{ij}(t) \frac{\partial^2 U}{\partial x_i \partial x_j} \tag{7}$$

with the dispersions

$$\alpha_i = \lim_{t\to 0} \frac{a_i}{t}$$

$$\beta_{ij} = \lim_{t\to 0} \frac{b_{ij}}{t}$$

belongs to a stochastic differential equation

$$\dot{x}_i = a_i x_i + b_{ij} w_j \tag{8}$$

with the noise processes $\underset{\sim}{W}(t)$. After a transformation of variables equ.(4) can be transformed into a form like (8) and the following equation can be deduced ($\xi = x_j$, $\eta = \dot{x}_j$), which looks like a backward equation

$$\frac{\partial U}{\partial t} = \eta \frac{\partial U}{\partial \xi} - (\nu_j^2 \xi + 2\mu_j \eta) \frac{\partial U}{\partial \eta} + D(t) \frac{\partial^2 U}{\partial \eta^2} \tag{9}$$

The coefficient $D(t)$ is evaluated from the autocorrelation of the excitation process

$$D(t) = \frac{1}{2} \int_{0}^{\infty} E\{\ddot{Z}(t)\ddot{Z}(t+\tau)\}\, d\tau \tag{10}$$

and gives a constant value for stationary processes $\ddot{Z}(t)$. The corresponding initial and boundary conditions are

$t = 0$	$U = 1$	(if a starting position within Γ was chosen)
$t > 0$	$U = 0$	for the boundary lines γ of the safe domain Γ and positive velocity

e.g. $\Gamma = \Gamma_d$

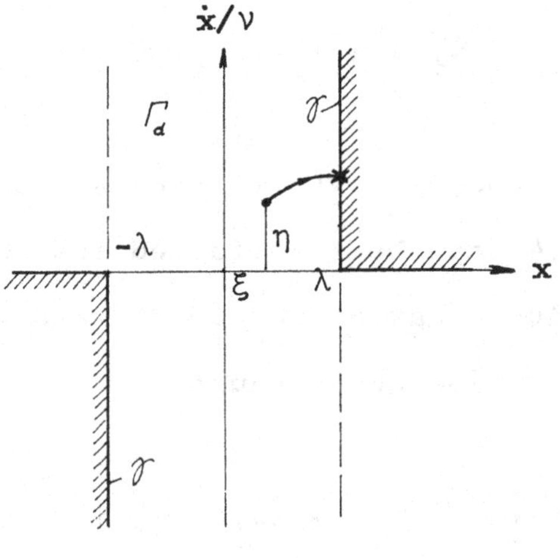

The dotted lines shall indicate, that no sample can reach them without having crossed the other ones before. After some doubts[5] it was shown, that the problem is well posed[6].

In the case of a stationary process $\ddot{Z}(t)$ the so-called Pontryagin's equation for the mean first passage time can be derived. After differentiating (9) with respect to time we get an analogous equation for the first-passage density $\vartheta(t;\lambda, \xi, \eta)$

$$\frac{\partial\vartheta}{\partial t} = \eta \frac{\partial\vartheta}{\partial\xi} - (\nu_j^2\xi + 2\mu_j\eta) \frac{\partial\vartheta}{\partial\eta} + D \frac{\partial^2\vartheta}{\partial\eta^2}$$

and after multiplication with t and partial integration we arrive at

$$- 1 = \eta \frac{\partial T}{\partial\xi} - (\nu_j^2\xi + 2\mu_j \eta) \frac{\partial T}{\partial\eta} + D \frac{\partial^2 T}{\partial\eta^2} \qquad (11)$$

with

$$T = \int_0^\infty t\vartheta(t;\lambda,\xi,\eta)\, dt$$

and the boundary condition $T = 0$ on γ. Nevertheless, exact solutions for equ.(11) are not known, but approximate results were obtained from Bolotin[7] by applying Galerkin's procedure. A comparison with exact results is only possible if x is determined from a first order differential equation driven by white noise, like

$$\dot{x} + \gamma x = n(t) \qquad \gamma = (-a, +a), \qquad E\{n(t)n(t+\tau)\} = c\delta(\tau) \tag{12}$$

where an exact solution is given by (Fig.3)

$$T(x) = -2\frac{\gamma a^2}{c}\int_{-1}^{x}\left[\exp\left(\frac{\gamma a \psi^2}{c}\right)\int_{0}^{\psi}\exp\left(-\frac{\gamma \eta^2 a}{c}\right)d\eta\right]d\psi \tag{13}$$

Because of possible exact solutions of the Pontryagin equation some authors tried to reduce the problem of the linear oscillator to a first order equation[3,8,9,10]. Rosenblueth suggested the introduction of a new variable

$$r = \sqrt{\nu_j^2 x^2 + 2\mu_j x\dot{x} + \dot{x}^2} \tag{14}$$

This approximation is reasonable, if the correlation time of the excitation process (which is zero for white noise and small for broadband processes) is much smaller than the natural period of the structure[10].

Another method to arrive at a one-dimensional Markov-process was suggested by Ariaratnam[8]. He set up the corresponding Fokker-Planck-equation

$$\frac{\partial p}{\partial t} = -\dot{x}\frac{\partial p}{\partial x} + (\nu_j^2 x + 2\mu\dot{x})\frac{\partial p}{\partial x} + D\frac{\partial^2 p}{\partial x^2} \tag{15}$$

converted it to polar coordinates a, φ and made use of an averaging technique analogous to that of Bogoliubov and

Mitropolsky[11]. If the fluctuation of the phase is neglected afterwards, a one-dimensional Markovprocess is obtained and the Pontryagin equation can be solved.

This procedure gives the following result for the mean first passage time and a safety domain Γ_e

$$T(a) = \frac{1}{2\mu_j}\left[\overline{Ei}(R^2) - \overline{Ei}(a^2) - \ln(\frac{R^2}{a^2})\right]$$
$$\overline{Ei}(x) = -\int_{-x}^{\infty} e^{-t}\frac{dt}{t} \quad x > 0 \tag{16}$$

The approximation leads to an upper bound for the first passage probability of the linear oscillator and a safe domain Γ_o or Γ_d. An extension of this procedure to the linear oscillator including the transient response has already been performed[12]. The solution is given in form of a series.

2.2 Numerical investigations

Because of the lack of exact solutions, numerical investigations have been performed very early. An extensive study of the first-passage probability of the linear oscillator driven be white noise was done by Crandall et al[13]. One drawback of simulation techniques lies in the fact, that the available information about crossing statistics decreases significantly with an increasing threshold level λ. Therefore, a great number of samples have to be created. The authors investigated the first passage probability using two approaches. First, they found from recurrence time statistics the first passage probability density (the recurrence time is the time interval between succeeding barrier crossings). Second, they investigated the time-behaviour of the probability mass in the phase plane and counted the loss of mass via the boundaries.

The figures 5 and 6 present some of their results.

For random initial conditions, the first passage probability density ϑ falls off significantly within the first cycles of response, After the transient response has died out, ϑ becomes independent of the initial conditions. It is observed that higher damping ratios or shorter correlation times yield higher first passage probabilities, if the threshold level λ is made dimensionless with the stationary rms-response of the structural response (equ.A3)

$$\sigma_{st}^2 = \frac{S_o}{4\mu_j \nu_j^2} \qquad b = \frac{\lambda}{\sigma_{st}} \tag{17}$$

where S_o is the constant spectral density of the excitation.

It can be assumed that, after having reached a steady-state response, the distribution of the first-passage probability density mass ϑ in the phase plane will remain invariant while its magnitude decays. Then the fraction of mass lost during an interval (indicating sample crossings of γ) will be proportional to the mass present at the beginning of the interval. These heuristic arguments yield the following expression for the tail of first-passage probability density curves[13]

$$\vartheta(t) = A\,\alpha\,e^{-\alpha t} \tag{18}$$

This formula was the subject of many investigations, and will be treated later on.

Another numerical procedure was used by Parkus and Zeman to solve a viscoelastic problem, where the excitation was modeled by filtered white noise[14]. The authors used an integral equation representation, that is analogous to equ.(7).

3. APPROXIMATE SOLUTIONS OF THE FIRST PASSAGE PROBLEM

3.1 Inclusion-Exclusion Series

Many authors start their approximations with the famous "inclusion-exclusion" series. In what follows, it will be derived.

Following Bartlett[15], we consider a discrete random process and denote by

$$\{e_r\} = P[S_r/S_o]$$

the probability, that conditional to the initial conditions the event S happens in r, regardless of all in-between states. S can be interpreted as the event of crossing the boundary γ of the safe region Γ with positive velocity. Using the further notation

$$\{\bar{e}_r\} = P[\text{not } S_r/S_o] = 1 - \{e_r\}$$

the probability of a <u>first</u> crossing event at n is given from

$$\{\bar{e}_1\ \bar{e}_2\ \bar{e}_3 \ldots \bar{e}_{n-1}\ e_n\} = \{(1-e_1)(1-e_2)\ldots(1-e_{n-1})e_n\} =$$
$$= \{e_n\} - \sum_{r=1}^{n-1}\{e_r e_n\} + \sum_{r=1}^{n-2}\ \sum_{s=r+1}^{n-1}\{e_r e_s e_n\} - + \ldots \tag{19}$$

Replacing the discrete time variable by a continuous one yields the series for a domain Γ_o or Γ_d

$$\vartheta(t;\lambda) = n_\lambda(t) - \int_0^t n_\lambda(t,s)ds + \int_0^t \int_u^t n_\lambda(t,s,u)ds\,du + \dots \tag{20}$$

where $n_\lambda(t_1,t_2,\dots t_n)$ represents the mean number of single or multiple crossings of the threshold level λ or $-\lambda$ from inside the safe domain Γ per unit time at the time instants $t_1,t_2\dots t_n$. The series leads to a system of lower and upper bounds because the crossing rates are essentially positive. It is noticed, that the response process $X(t)$ is assumed not to change its characteristic features after having exceeded the boundary γ once.

The mean crossing rates n_λ are computed from the integrals

$$n_\lambda(t) = \int_0^\infty \dot{x}p(\lambda,\dot{x};t)d\dot{x} \quad \text{for } \Gamma_o$$

$$\text{or} \quad 2\int_0^\infty \dot{x}p(\lambda,\dot{x};t)d\dot{x} \quad \text{for } \Gamma_d \tag{21}$$

and the multiple crossing rates, consequently, follow from

$$n_\lambda(t_1,t_2) = \int_0^\infty d\dot{x}_1 \int_0^\infty p(\lambda,\dot{x}_1,\lambda,\dot{x}_2;\ t_1,t_2)d\dot{x}_2 \quad \text{for } \Gamma_o \tag{22}$$

Substitution of the Gaussian joint probability density of the structural response process $X(t)$ and its time derivative $\dot{X}(t)$

$$p(x,\dot{x};t) = \frac{1}{2\pi\sigma_1\sigma_2\sqrt{1-\rho_{12}^2}} \exp - \left[\frac{x^2\sigma_2^2 - 2\rho_{12}x\dot{x}\sigma_1\sigma_2 + \dot{x}^2\sigma_1^2}{2\sigma_1^2\sigma_2^2(1-\rho_{12}^2)}\right] \tag{23}$$

with the variance functions (given in the appendix)

$$\sigma_1^2 = E\{X^2(t)\}$$
$$\sigma_2^2 = E\{\dot{X}^2(t)\} \tag{24a}$$

and their correlation coefficient

$$\rho_{12} = \frac{E\{X(t)\dot{X}(t)\}}{\sigma_1\sigma_2} \tag{24b}$$

into equ.(21) yields the following expression for the mean rate of upward crossings of the level λ

$$n_\lambda^+(t) = n_o^+(t)e^{-\lambda^2/2\sigma_1^2}\left[e^{v_1^2} + v_1\sqrt{\pi}(1 + \operatorname{erf}(v_1))\right] \tag{25}$$

where erf is the widely known error function

$$\operatorname{erf}(x) = \frac{2}{\sqrt{\pi}}\int_o^x e^{-t^2}dt \tag{26}$$

The mean rate of crossing the zero axis with positive velocity is given from

$$n_o^+(t) = \frac{\sigma_2}{2\pi\sigma_1}\sqrt{1-\rho_{12}^2} \tag{27}$$

and the abbreviation

$$v_1 = \frac{\lambda\rho_{12}}{\sigma_1\sqrt{2(1-\rho_{12}^2)}} \tag{28}$$

vanishes for stationary processes, because the process and its time derivative are uncorrelated in this case. Hence, a simpler expression instead of (25) is obtained

$$n_\lambda^+ = \frac{\sigma_2}{2\pi\sigma_1}\exp\left(-\frac{\lambda^2}{2\sigma_1^2}\right) \tag{25'}$$

Because of the symmetric properties of X(t) and symmetric boundaries

$$n_\lambda^+ = n_{-\lambda}^-$$

holds. Fig.7 shows a result of the time-behaviour of $n_\lambda^+(t)$ for the FNW-excitation.

It can be concluded that equ.(20) is a conceptually relatively simple approach, but leads to computational difficulties because of the multiple integration of the multidimensional probability density in equ.(22).

Series equ.(20) was used by ROBERTS to compute the first-passage probability density ϑ of a linear oscillator (equ.4) where the excitation process was modeled by stationary[16,17] or nonstationary[18] white noise. Roberts truncated the series expansion after the third term. It is not possible to give a closed-form solution of the second and third term like equ.(25) for the first term, and numerical integration has to be performed, therefore.

3.2 Upper and lower bound I

To overcome this difficulty, further approximations were suggested. Consideration of only the first term in the series equ.(20) yields an upper bound, that can be evaluated easily

$$\vartheta(t;\lambda) < n_\lambda(t) = \begin{cases} n_\lambda^+(t) & \text{for } \Gamma_o \\ n_\lambda^+ + n_{-\lambda}^- = 2n_\lambda^+ & \text{for } \Gamma_d \end{cases} \tag{29}$$

A similar result for the first-passage probability was obtained by Shinozuka[19] from different considerations.

$$H(t;\lambda) \leq \int_0^t n_\lambda(\tau)d\tau - \int_{-\infty}^{-\lambda} dx_1 \int_\lambda^\infty p(x_1,x_2;\tau_1,\tau_2)dx_2 \tag{30}$$

The second term expresses the probability, that the process

is outside of the safe domain Γ_d at two different time instants within the interval (0,t), that have to be found from optimization considerations. For narrowband processes, like the response of a linear, slightly damped oscillator to a broadband process, the instants τ_1, τ_2 are chosen optimal with[20]

$$\tau_1 = \tau_2 - \frac{\pi}{\nu_j} \tag{31}$$

where the autocorrelation coefficient has a local minimum.

For stationary processes the mean crossing rate (25') is time-independent, and the upper bound is, therefore, a linear function of t. In other words, equation (30) is not reasonable for very large t, where the upper bound tends to infinity.

For nonstationary processes, like white noise multiplied with a deterministic envelope function, that decays to zero for large time instants, the mean crossing rate n_λ is a time-dependent function, too (Fig.7). Because of the dieing-out of the excitation, n_λ will also decay and the integral in equ.(30) will reach a stationary value. The second term in equ.(30) can be maximized, if τ_2 is chosen as that specific moment, when $\sigma_1(t)$ reaches its maximum value σ_1^*, and equ.(31) is taken into account. Obviously, upper bounds greater than 1 are trivial results.

The advantage of an upper bound equ.(30), which will

be referred to as upper bound I in what follows, is its relatively simple form and the fact, that not much information need be known about the response process. In many cases, the second term, for whose computation the autocorrelation function of X(t) is needed, has negligable influence (about 15 % for relatively small λ-values, and less than 1 % for higher ones).

The variance functions of the structural response X(t) are easier to compute than the autocorrelation function.

Shinozuka has added to the upper bound I a lower bound (it will be named lower bound I)

$$H(t;\lambda) > 2 \int_{\lambda}^{\infty} p(x;t^*)dx \tag{32}$$

and for Gaussian processes

$$H(t;\lambda) > 1 - \operatorname{erf}\left(\frac{\lambda}{\sigma_1^* \sqrt{2}}\right) \tag{33}$$

Herein t^* is that time instant, where the rms function of the process X(t) reaches its maximum value. For stationary processes this will be the steady-state response level.

The distance between these two bounds depends on the parameters of the system and its excitation. Generally speaking, the bounds are closer for soft systems with relatively low fundamental eigenfrequencies and for systems with higher damping values (Fig.8). Unfortunately, we cannot assume these properties to hold in general. On the contrary, we have to deal usually with light damped, relatively rigid, structures. In any case, the lower and upper bounds can simply be evaluated and possess the pleasant property, that they enable definitively conservative or non-conservative design.

3.3 Poisson Process Approximations

A question of major interest in the application of the inclusion-exclusion series equ.(20) is that of its convergence. It is known[17] that for large values of t the series diverges. Therefore, a transformation has to be performed

$$\vartheta = A \exp \sum_{s=1}^{\infty} \frac{(-1)^s}{s!} \int_0^t \cdots \int_0^t g_s(\tau_1, \ldots \tau_s) d\tau_1 \ldots d\tau_s \qquad (34)$$

whereas the cumulant functions $g_s(\tau_1 \ldots \tau_s)$ can be obtained from the crossing rates by expanding the right hand side of equ.(34) and comparing it with equ.(20). Thus the following relations are obtained

$$g_1(t) = n_\lambda(t)$$

$$g_2(t_1, t_2) = n_\lambda(t_1, t_2) - n_\lambda(t_1) n_\lambda(t_2)$$

$$g_3(t_1,t_2,t_3) = n_\lambda(t_1,t_2,t_3) - n_\lambda(t_1)n_\lambda(t_2,t_3) -$$

$$- n_\lambda(t_2)n_\lambda(t_1,t_3) - n_\lambda(t_3)n_\lambda(t_1,t_2)$$

$$+ 2n_\lambda(t_1)n_\lambda(t_2)n_\lambda(t_3)$$

The cumulant functions characterise the correlation of arrival rates at different instants of time. If at least one crossing rate is uncorrelated with all others, then $g_s(t_1,\ldots t_n) = 0$. Thus we arrive immediately at the exponential distribution ϑ_p, if we assume all crossings of the tolerable threshold level to be statistically independent. They form a nonhomogeneous Poisson process with the mean arrival rate n_λ. The first passage probability density has an exponential distribution

$$\vartheta_p(t;\lambda) = n_\lambda(t)e^{-\int_0^t n_\lambda(\tau)d\tau} \tag{35}$$

$$H_p(t;\lambda) = 1 - e^{-\int_0^t n_\lambda(\tau)d\tau} \tag{36}$$

In the stationary case the following equations

$$\vartheta_p(t;\lambda) = n_\lambda e^{-n_\lambda t} \tag{35'}$$

$$H_p(t;\lambda) = 1 - e^{-n_\lambda t} \tag{36'}$$

yield a mean first-passage time n_λ^{-1}.

In the limit $t \to \infty$ $H(t;\lambda)$ tends to one. That means that almost all samples have already exceeded the threshold level at least once. It is remarkable that this limit is independent of λ.

It is pointed out that the exponent in equ.(35,36) is just the upper bound I. As was already stated, the integrals reach some stationary value and the first-passage probability equ.(35) is, therefore, always less than one, even if t goes to infinity. This is a remarkable difference to the stationary case.

If the level λ is very high, n_λ will be very small and the upper bound I and the result of equ.(36) will coincide. Moreover, it can be shown that, in the limit $\lambda \to \infty$, the Poisson assumption tends to the exact solution for both stationary[21] and nonstationary[18] processes.

These arguments explain the reason for the fact that the exponential distribution is widely used and can be referred to as a "central approximation", because many other solutions are measured by comparison with this simple assumption.

As an example, Rice and Beer[22] suggested to simplify the inclusion-exclusion series by assuming the properties of a renewal process for the barrier crossings. That means that the event of an upward crossing depends only on the last prior upward crossing. By the help of this assumption all multiple crossing probabilities of order greater than two can be computed from the first two ones. The numerical results given in[22] for a stationary problem indicate, that the renewal process approximation tends to the Poisson approximation for higher threshold levels ($\lambda = 2\text{-}3\sigma_{st}$) and yields smaller first-passage densities for lower λ-values.

In the case of stationary processes one is usually interested in the limiting decay rate. According to Cramér and Leadbetter[21] it must tend to n_λ for increasing λ-values. Therefore we are motivated to write

$$H(t;\lambda) = 1 - \exp\left[- n_\lambda t(1 - A_2 + A_3 - \dots)\right] \tag{37}$$

Roberts[17] has suggested some approximations to compute the series equ.(37) from A_2 or A_2 and A_3 only. He either assumed the exponent to be a geometric series

$$H_G(t;\lambda) = 1 - \exp\left[- n_\lambda t(.5 + \frac{1-A_2}{4} + \frac{1-2A_2+A_3}{8})\right] \tag{38}$$

or used a nonlinear Shanks' transformation

$$H_S(t;\lambda) = 1 - \exp\left[- n_\lambda t(1 - \frac{A_2}{1+A_3/A_2})\right] \tag{39}$$

He found good agreement, particularly for the second assumption, between these approximations and simulation results.

A careful study of all of these corrections of the exponential distribution ϑ_p reveals, that the Poisson assumption leads to conservative results for narrowband processes. On the other hand, however, nonconservative results are obtained for broadband processes. The conservativism can be removed for narrowband processes by introducing an envelope process. The envelope definition must be appropriate for both stationary and nonstationary processes.

3.4 Envelope definition

Following Priestly[23], each sample of a real process with evolutionary power spectrum has a spectral representation

$$X(t) = \frac{1}{\sqrt{2\pi}} \int_{-\infty}^{\infty} M(t,\omega) e^{i\omega t} dZ(\omega) = \frac{1}{\sqrt{2\pi}} \int_{-\infty}^{\infty} \bar{M}(t,\omega) e^{-i\omega t} d\bar{Z}(\omega) \tag{40}$$

$M(t,\omega)$ is a complex modulating function that reduces to 1 for stationary processes, and the random increments $dZ(\omega)$ have the same nice property (a bar shall denote the complex conjugate)

$$E\left\{dZ(\omega_1)d\bar{Z}(\omega_2)\right\} = S(\omega_1)\delta(\omega_1-\omega_2)d\omega_1\, d\omega_2 \tag{41}$$

as in the stationary case. $S(\omega_1)$ is the spectral density of the corresponding stationary process with the corresponding spectral representation, which is due to Wiener[32]

$$X^{+}(t) = \frac{1}{\sqrt{2\pi}} \int_{-\infty}^{\infty} e^{i\omega t} dZ(\omega)$$

The process X(t) is formed by sine-and cosinewaves with time-varying amplitudes, and equ.(34) is valid, if the Fouriertransform of $M(t,\omega)$

$$M(t,\omega) = \int_{-\infty}^{\infty} e^{itu} \, dH(u)$$

has an absolute maximum in $u = 0$. This assumption is generally justified, when the nonstationary response of the structure under earthquake acceleration is considered. The modulating functions for various examples of the excitation process are given in the appendix (equ.A11, A16).

Generalising a definition from Cramér and Leadbetter[21] we define a real process Y(t) by

$$Y(t) = \frac{1}{\sqrt{2\pi}} \int_{-\infty}^{\infty} M(t,\omega) g(\omega) e^{i\omega t} dZ(\omega) \tag{42}$$

with the gain

$$g(\omega) = - i \,\mathrm{sgn}(\omega) \tag{43}$$

This process in orthogonal to X(t), as can be seen, if the correlation functions are considered $(s > t)$

$$R_{xx}(t,s) = E\left\{X(t)X(s)\right\} = \tag{44}$$

$$= \frac{1}{\pi}\int_0^{\infty}\bar{M}(t,\omega)M(s,\omega)\cos\,\omega(s-t)S(\omega)d\omega$$

The same correlationfunction is found for the process Y(t)

$$R_{yy}(t,s) = E\left\{Y(t)Y(s)\right\} = R_{xx}(t,s)$$

The crosscorrelation function

$$R_{xy}(t,s) = E\left\{X(t)Y(s)\right\} =$$

$$= \frac{1}{2\pi}\int_0^{\infty}\bar{M}(t,\omega)M(s,\omega)e^{-i\omega(s-t)}(-i)S(\omega)d\omega +$$

$$+ \frac{1}{2\pi}\int_{-\infty}^{0}\bar{M}(t,\omega)M(s,\omega)e^{i\omega(s-t)}(i)S(\omega)d\omega =$$

$$= \frac{1}{\pi}\int_0^{\infty}\bar{M}(t,\omega)M(s,\omega)\sin\,\omega(s-t)S(\omega)d\omega \tag{45}$$

vanishes for $s = t$, as was stated above.

Considering the components of the frequency composition of the real processes X(t)

$$M(t,\omega) = \left|M(t,\omega)\right|\,e^{i\gamma(t,\omega)}$$

$$M(t,\omega)e^{i\omega t}dZ(\omega) = |M|e^{i\gamma+i\omega t}(dU + idV) =$$
$$= |M|(\cos(\omega t + \gamma)dU - \sin(\omega t + \gamma)dV)$$

and of Y(t)

$$M(t,\omega)e^{i\omega t}\, g(\omega)\, dZ(\omega) = |M|(\sin(\omega t + \gamma)dU + \cos(\omega t + \gamma)dV)$$

it is observed, that the elementary components of Y(t) are out of phase by 90 degrees with respect to the components of X(t). Thus, the result of equ.(45) is not surprising.

The envelope process is defined with

$$A(t) = \sqrt{X^2(t) + Y^2(t)} \qquad (46)$$

It is obvious that

$$A(t) \geq X(t)$$

and

$$A(t) = X(t) \quad \text{if} \quad Y(t) = 0$$

holds.

This definition of the envelope process can be shown to be equivalent to that of Rice[24]

$$X(t) = A(t)\cos(\omega_m t + \phi(t)) \qquad (47)$$

if the midfrequency ω_m in his formula for stationary processes is put equal to zero

$$X(t) = A(t)\cos\phi(t)$$
$$Y(t) = A(t)\sin\phi(t)$$

Equ.(46) can be applied to both broadband and narrowband processes, as the Figures 2 and 3 in the chapter of Prof. Shinozuka will illustrate.

Although not being strictly necessary the existence of an oscillatory process has been assumed throughout the derivation of the envelope process. For instance, the crosscorrelation function $R_{xy}(t,s)$ could also be evaluated from the two-sided spectral density $S(\omega_1,\omega_2)$[35] with the dummy parameters ω_1,ω_2

$$R_{xy}(t,s) = \frac{1}{2\pi^2} \int_{-\infty}^{\infty} e^{i\omega_1 t} \int_{0}^{\infty} \sin \omega_2 s \; S(\omega_1,\omega_2)\, d\omega_1 \, d\omega_2$$

However, in applications, particularly for multi-degree-of-freedom systems, the double integration is much more complicated than the evaluation of equ.(45). In addition, the description of the process by its evolutionary spectral density

$$S_t(\omega) = |M(t,\omega)|^2 S(\omega)$$

retains some physical interpretations of the spectra.

The probability distribution of the envelope process and the phase are obtained by inserting the Gaussian probability distributions of X(t) and Y(t). The envelope process A(t) is Rayleigh distributed (Fig.9).

$$p(a;t) = \frac{a}{\sigma_1^2(t)} \exp\left(- \frac{a^2}{\sigma_1^2(t)}\right) \tag{48}$$

and the phase is equally distributed in the interval $(0, 2\pi)$.

Any other n-dimensional probability density of the envelope process can be derived, in principle, from the 2n-dimensional Gaussian probability distribution. In what follows, this shall be demonstrated with the joint probability density of $A(t)$ at two different time instants t_1 and t_2.

We start with the fourdimensional Gaussian probability density for the processes X and Y at two different time instants and introduce the new variables

$$x_1 = X(t_1) \qquad x_2 = X(t_2) \qquad x_3 = Y(t_1) \qquad x_4 = Y(t_2)$$

$$p(x_1, x_2, x_3, x_4) = \frac{1}{4\pi^2\sqrt{|M|}} \exp\left(- u_{ik}\, x_i\, x_k\right)$$

The corresponding covariance matrix M has the form

$$\begin{vmatrix} \sigma_1^2 & 0 & \sigma_1\sigma_2 k & \sigma_1\sigma_2 r \\ 0 & \sigma_1^2 & -\sigma_1\sigma_2 r & \sigma_1\sigma_2 k \\ \sigma_1\sigma_2 k & -\sigma_1\sigma_2 r & \sigma_2^2 & 0 \\ \sigma_1\sigma_2 r & \sigma_1\sigma_2 k & 0 & \sigma_2^2 \end{vmatrix} \tag{49}$$

with the determinant $|M| = \sigma_1^4\sigma_2^4(1-k^2-r^2)$, and with u_{ik} as the elements of M^{-1}. Herein, σ_1 and σ_2 denote the rms-values of $X(t_1)$ or $X(t_2)$, k and r denote the auto- respective crosscorrelation coefficient of X(t) and Y(t). After a transformation of variables

$$x_i = a_i \cos \varphi_i \qquad i = 1,2$$

$$x_i = a_i \sin \varphi_i \qquad i = 3,4$$

and integration over the phase angles from 0 to 2π, the following density is obtained[4]

$$p(a_1,a_2) = \frac{a_1 a_2}{\sigma_1^2\sigma_2^2(1-k^2-r^2)} J_o\left(\frac{a_1 a_2\sqrt{k^2+r^2}}{\sigma_1\sigma_2(1-k^2-r^2)}\right)$$

$$\exp - \left[\frac{a_1^2\sigma_2^2 + a_2^2\sigma_1^2}{2\sigma_1^2\sigma_2^2(1-k^2-r^2)}\right] \tag{50}$$

Herein, J_o is the modified Bessel function of zero order and complex argument[26]

$$J_o(q) = \frac{1}{2\pi}\int_o^{2\pi} \exp(q \cos \gamma)d\gamma \tag{51}$$

If we are interested in the joint probability density of the envelope process and its time derivative at the same moment, we may obtain this result from (50) as the limit

$$p(a_1,\dot{a}_1;t) = \lim_{\tau\to 0} p(a_1,a_2;t,t+\tau)$$

If the time instants t_1 and t_2 in equ.(50) come closer, the

Bessel function tends to infinity, and the exponential function tends to zero. Thus, an indefinite expression is obtained. We expand, therefore, both arguments into a series with $\tau = t_2 - t_1$, and make use of the expansion of the Bessel function for large arguments[26]

$$J_o(q) = \frac{e^q}{\sqrt{2\pi q}}(1 + \ldots) \tag{52}$$

$$\begin{aligned}
\sigma_1\sigma_2 k &= \sigma_1^2(1 + k'\tau + k''\frac{\tau^2}{2} + \ldots) \\
\sigma_1\sigma_2 r &= \sigma_1^2(0 + r'\tau + r''\frac{\tau^2}{2} + \ldots) \\
\sigma_2^2 &= \sigma_1^2(1 + 2k'\tau + s''\frac{\tau^2}{2} + \ldots)
\end{aligned} \tag{53}$$

All abbreviations are given in the appendix (equ.A18 - A22).

In the limit $\tau \rightarrow 0$ we obtain

$$p(a,\dot{a};t) = \frac{a}{\sigma_1^3\sqrt{2\pi\Delta}} \exp\left[-\frac{a^2}{2\sigma_1^2} - \frac{(\dot{a}-ak')^2}{a\sigma_1^2\Delta}\right] \tag{54}$$

In the stationary case this density can be split into a Gaussian density for the envelope time derivative and a Rayleigh distribution for the envelope because k' vanishes (equ.A18)

$$p(a,\dot{a}) = \frac{1}{\sigma_1\sqrt{2\pi\Delta}} \exp\left(-\frac{\dot{a}^2}{2\sigma_1^2\Delta}\right) \cdot \frac{a}{\sigma_1^2} \exp\left(-\frac{a^2}{2\sigma_1^2}\right) \tag{54'}$$

$$= p(\dot{a}) \cdot p(a)$$

The envelope and its time derivative are statistically independent in this case.
Hence, for a stationary process, the correlation coefficients (equ.24b)

$$\rho_{x\dot{x}} = \frac{E\{X(t)\dot{X}(t)\}}{\sigma_1\sigma_2} = \rho_{12}$$

and

$$\rho_{a\dot{a}} = \frac{E\{(A(t) - \sqrt{\frac{\pi}{2}}\,\sigma_1)\dot{A}(t)\}}{\sqrt{(1 - \frac{\pi}{2})\sigma_1^2\,\sigma_1^2\Delta}} = \frac{k'}{2\sqrt{(2 - \frac{\pi}{2})\Delta}}$$

both vanish. The nonstationary character of any process can, therefore, be evalued by the help of these correlation coefficients.

The bandwidth Δ of the structural response is given from the expression

$$\Delta = \omega_2 - \omega_1^2$$

$$= \frac{s''}{2} - k'' - (k'^2 + r'^2) \qquad (55)$$

Using the evolutionary spectra the modified first and second moments are computed from the integrals

$$\omega_1 = \frac{1}{\pi\sigma_1^2} \int_0^\infty \left[\underline{\bar{M}(t,\omega) \frac{\partial M(t,\omega)}{\partial t}} + \omega M(t,\omega)\bar{M}(t,\omega) \right] S(\omega)d\omega$$

$$\omega_2 = \frac{1}{\pi\sigma_1^2} \int_0^\infty \left[\underline{\frac{\partial \bar{M}(t,\omega)}{\partial t} \frac{\partial M(t,\omega)}{\partial t}} + \omega^2 \bar{M}(t,\omega)M(t,\omega) \right] S(\omega)d\omega \tag{56}$$

where the underlined terms vanish in the stationary case because of $M(t,\omega) \equiv 1$.

The specific bandwidth introduced by Vanmarcke[27] has to be written as

$$q_v = \sqrt{\frac{\Delta}{\omega_2}}$$

and varies between 0 and 1.

Fig.10 presents an illustration of the evolutionary spectral density for the time derivative of the structural response process of a shearbeam, that was modeled by a three-degrees-of-freedom system with structural damping ($\zeta_j \nu_j$ = const.). Resonance peaks and the time-varying character of the spectrum are obvious.

3.5 Approximations based on the envelope process

First of all, improved upper and lower bounds can be derived by the help of the envelope process. Narrowband processes tend to exceed a certain threshold level λ in clumps. Among various definitions of the mean clumpsize e.g.[33], the one proposed by Vanmarcke[27] seems favourable. It reflects the fact that on the one hand, crossings occur in clusters, and on the other hand, envelope crossings are only to be taken into account, if they are followed by crossings of the process X(t) itself.

$$E\{cs\}_v = \frac{1}{1-e^{-\alpha_d}} \tag{57}$$

$$\alpha_d = \frac{N_\lambda}{2n_o^+} e^{\lambda^2/2\sigma_1^2} \qquad \Gamma = \Gamma_d$$

These equations hold for stationary processes and for those nonstationary processes, where N_λ, n_o^+, σ_1 do not change significantly in time. Investigation of the response of the linear oscillator to various inputs (NW, FNW) leads to the result that, particularly in the period of increase, a significant nonstationary character can be observed (Fig.22). On the other hand, the response level will be relatively small during that period. In many cases the formulas for the stationary case can, therefore, be applied without significant error. Otherwise, numerical integration must be performed.

The mean rate of envelope upward crossings N_λ follows from evaluating the integral.

$$N_\lambda(t) = \int_0^\infty \dot{a}p(\lambda,\dot{a};t)d\dot{a} =$$

$$= \frac{\lambda}{\sigma_1}\sqrt{\frac{\Delta}{2\pi}}\, e^{-\lambda^2/2\sigma_1^2}\left[e^{-v_2^2} + v_2\sqrt{\pi}(1 + \mathrm{erf}(v_2))\right]$$

$$v_2 = \frac{\lambda}{\sigma_1}\frac{k'}{\sqrt{2\Delta}} \tag{58}$$

and reduces in the stationary case to

$$N_\lambda = \frac{\lambda}{\sigma_1}\sqrt{\frac{\Delta}{2\pi}}\, e^{-\lambda^2/2\sigma_1^2} \tag{54'}$$

$$= 2n_\lambda^+ \frac{\lambda}{\sigma_2}\sqrt{\frac{\pi\Delta}{2}}$$

Fig.11 shows a result of $E\{cs\}_v$ and $N_\lambda/2n_\lambda^+$ depending on λ for a specific example.

Similarily to equ.(32), a lower bound II is obtained[4] from

$$H(t;\lambda) \geq \int_\lambda^\infty p(a;t)da = e^{-\lambda^2/2\sigma_1^{*2}} \tag{59}$$

Taking into account the mean clumpsize, an upper bound II is found from the inequality

$$H(t;\lambda) \leq \int_0^t \frac{n_\lambda(\tau)}{E\{cs(\tau)\}}d\tau = \int_0^t \underbrace{2n_\lambda^+(1-e^{-\alpha_d})}_{q_\lambda(\tau)}\, d\tau \tag{60}$$

The integrand q_λ is sometimes referred to as mean number of qualified envelope (upward) crossings. Moreover, one can assume, that these crossings are statistically independent and, therefore, form a Poisson process[27], thus arriving at

$$H(t;\lambda) = 1 - e^{-\int_0^t q_\lambda(\tau)d\tau} \qquad (61)$$

Figure 12 illustrates the results for the first-passage probability for a flexural beam model. The structure was found to possess three significant degrees of freedom. The excitation was modeled with a FNW-acceleration process according to[38] (see appendix). For the selected parameters the maximum rms-value of the excitation was 0,12 g. The fundamental frequency of the structure was 0,84 Hz and the damping ratio in the first mode was 2 %. According to equation (3) a maximum rms-value of the lateral end deflection $\sigma_1^* = 0{,}0125\ l$ was obtained.

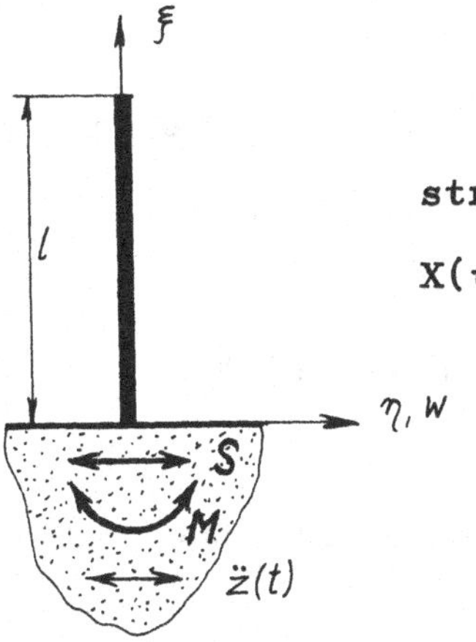

structural response

$X(t) = w(\xi = 1, t)$

The lower bound II gives an overall improvement over the lower bound I. The improvement of the upper bound II can be shown to be more significant for one-degree-of-freedom systems, where the bandwidth of the corresponding spectra is smaller (Fig.13 presents a result for the same excitation and same structural model, as in Fig.12, if only the first mode of vibration is taken into account). The diagrams 12 and 13 can be interpreted in the following two ways.

If we assume a maximum beam end deflection of 4 % of the beam length to be tolerable, the following result for the reliability U of this structure (see Fig.12) is obtained

$$\frac{\lambda}{\sigma_1^*} = 3,2, \qquad \Gamma = \Gamma_d$$

$0,6\ \% < H < 3,9\ \%$ bounds for the first-passage probability

$99,4\ \% > U > 96,1\ \%$ bounds for the reliability

In this case the maximum rms value of the base shear force S was computed from

$$\sigma_s^* = \max_t \left[\sum_{j=1}^{n} (b_j\ EJ\ f_j'''(0) x_j(t))^2 \right]^{1/2}$$

and results in 0,605 G, where G represents the total weight of the column.

As a second example, let us assume a FNW excitation, that yields a maximum rms-value $\sigma^*_{\ddot{z}}$ of the excitation process of 0,05 g. The peak values of the corresponding samples are within the range from 0,08 to 0,15 g. The structural model will be the same as before, but only the first mode is taken into account (Fig.13). If we select a value of 99 % for the reliability, that has to be guaranteed, we conclude from Fig.13 that the corresponding tolerable structural response level can be bounded as follows

$$3{,}05\sigma^* < \lambda < 3{,}52\sigma^*, \qquad \sigma^* = \max_t \int_0^t h(t-\tau)\ddot{z}(\tau)d\tau$$

Considering the base shear force, we have, therefore, to design the structure in such a way, that a maximum shear force of $S_{max} = 1{,}142\ \sigma^*_{\ddot{z}} \cdot G \cdot 3{,}52 = 0{,}20\ G$ can be submitted. As a comparative result, a base shear force according to the Uniform Building Code

$$S = \frac{0{,}05}{\sqrt[3]{T}}\ G = \frac{0{,}05}{\sqrt[3]{1{,}192}}\ G = 0{,}185\ G$$

is obtained.

A comparison of the results for the upper and lower bounds with other investigations (e.g. point approximations) or simulation results reveals that the solution for the first-passage probability will be closer to the upper bound.

As already mentioned, the upper bound II becomes the exact solution for very large barrier levels.

The Figures 12, 13 are completed by an upper bound III, that was derived by the author by summarizing over all those envelope crossings that remain below λ within the safe region one cycle earlier[4]. The required probability distribution

$$p(a_1, a_2, \dot{a}_2;\ \ t_1, t_2, t_2)$$

must be evaluated by numerical integration. Upper bound III yields better results than upper bound II for small barrier levels only. Comparison with simulation results indicate that the upper bounds II and III could be approximately combined. For further details see[4,28].

3.6 Point approximations

These approximations discretise the continuous time variable t. The properties of a narrowband-process can be used to construct a process consisting of all extrema of X(t). Shinozuka and Yang developed this concept for stationary processes[29]. Recently, it was applied to nonstationary processes, too[30]. The authors use for the statistics of the point process those of the envelope process. That means, that the probability distribution function of an extremal point η at $t = nT_o/2$, with T_o being the apparent period, is approximated by

$$F_\eta(y) = \int_o^y p(a;t)da = 1 - e^{-\lambda^2/2\sigma_1^2} \quad (62)$$

and the joint density function of two extremal points at different numbers of cycles follows from

$$p(\eta_1,\eta_2;\ n_1T_o/2,\ n_2T_o/2) \doteq p(a_1,a_2;\ t_1,t_2) \quad (63)$$

by inserting equ.(50)

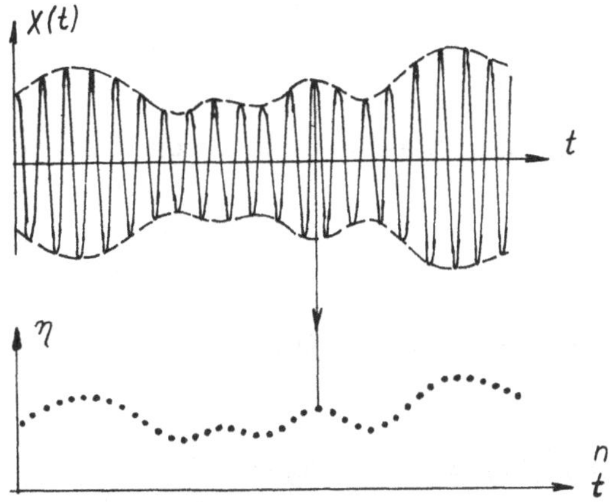

extreme point process

The inclusion-exclusion series equ.(20) can be applied now, if we replace the crossing rates by the probabilities that the point process remains above λ.

$$n_\lambda(t) \longrightarrow e^{-\lambda^2/2\sigma_1^2(t)} = \bar{n}_\lambda(t)$$

$$n_\lambda(t_1,t_2) \longrightarrow \int_\lambda^\infty \int_\lambda^\infty p(a_1,a_2;t_1,t_2)da_1 da_2 = \bar{n}_\lambda(t_1,t_2) \qquad (64)$$

Similarily, the corresponding cumulant functions in equ.(34) have to be replaced

$$g_1(t_1) = \bar{n}_\lambda(t_1)$$

$$g_2(t_1,t_2) = \bar{n}_\lambda(t_1,t_2) - \bar{n}_\lambda(t_1)\bar{n}_\lambda(t_2)$$

and so on

Shinozuka and Yang derive results from the following approximations

1) assumption of statistical independence of the random points

$$\bar{H}_p(NT_o/2) \doteq 1 - \exp\left(- \sum_{k=1}^{N} \bar{n}_\lambda(k)\right) \qquad (65)$$

2) for Gaussian processes the cumulant functions of order greater than two vanish. An approximation, which neglects all cumulants of order higher than two in equ.(34), can be interpreted as underlying a pseudo-Gaussian process.

$$\bar{H}(NT_o/2) \doteq 1 - \exp\left[-\sum_{k=1}^{N} \bar{n}_\lambda(k) + \frac{1}{2}\sum_{k_1=1}^{N}\sum_{k_2=1}^{N}(\bar{n}_\lambda(k_1,k_2) - \bar{n}_\lambda(k_1)\bar{n}_\lambda(k_2))\right] \tag{66}$$

3) $\eta(k)$ shall be Markovian. This assumption correlates the random point $\eta(k)$ with $\eta(k-1)$ only, and yields

$$\bar{H}_M(NT_o/2) = 1 - \exp\left[-\sum_{k=1}^{N} \bar{n}_{\lambda,M}(k)\right] \tag{67}$$

$$\bar{n}_{\lambda,M}(k) = P\left[\eta(k) = \lambda/\eta(k-1) < \lambda\right] = \frac{\int_0^\lambda da_1 \int_\lambda^\infty p(a_1,a_2;\ k-1,k)\,da_2}{1 - \bar{n}_\lambda(k-1)}$$

4) lower bounds can be derived from

$$\bar{H}(NT_o/2) \geq Q^T\pi^{-1}Q \geq L \tag{68}$$

where Q is an N-component vector with the elements $\bar{n}_\lambda(k)$, π is an N x N matrix with the elements

$$\pi_{kl} = \bar{n}_\lambda\ (k,l) \qquad \pi_{kk} = \bar{n}_\lambda(k)$$

and Q^T is the transpose of Q and

$$L = (\sum \bar{n}_\lambda(k))^2 / \sum_{k_1=1}^{N} \sum_{k_2=1}^{N} \bar{n}_\lambda(k_1,k_2)$$

An effective algorithm to compute the integrals equ.(63) readily is given by Yang[30].

These approximations are plotted in Fig.14 against the threshold level λ for a linear oscillator excited by the NW-model[30]. Additionely, the lower and upper bound II (equ.59 and 60) and Vanmarcke's approximation (equ.61) are shown.

As can be seen from Fig.14, the assumed Markov-property solution and the lower bound equ.(68) come very close. Vanmarcke's approximation lies above the Markov-approximation equ.(67), but is easier to compute. The upper bound II coincides with equ.(61) for threshold levels greater than 2,7 b. The Poisson assumption equ.(65) gives the worst result.

It must be noted, that one of the fundamental assumptions of the point process approximation is given by the fact, that the apparent period of the structure is nearly time-invariant. Fig.15 shows the time-behaviour of the frequency $\sqrt{\omega_2}$(equ.56) for a single-degree-of-freedom system and FNW-excitation. After the build-up period, ranging from 0 to about 5 seconds, the frequency is approximately constant.

A further, and indeed very severe restriction is due to the fact, that point approximation cannot be applied to multi-degree-of-freedom systems, where more than one apparent frequency exist. The corresponding spectrum becomes broader and narrowband approximations are, therefore, not useful in that case.

Another point process approximation was proposed by Lin[31]. Each barrier upward crossing is considered as a random point in time. Thus the time interval between the occurence of these random points is depending on the threshold level, and points occur sometimes in clumps for narrowband processes. The corresponding cumulant functions for the first two terms in the series (34) are evaluated from the statistics of the Gaussian process. The computation is more time-consuming and the approximation suggested by Shinozuka and Yang seems more appropriate.

Finally, a further comparison between various approximations for the first-passage probability is given in Fig.16.

A linear damped oscillator excited by a NW-process serves as an example in this case. All results are compared with the fundamental Poisson process assumption $H_p(\lambda)$(equ.35). B_1, B_2 and B_3 are the first three bounds obtained by Roberts[18] from the inclusion-exclusion series. For relatively low barrier levels the series diverges. For higher b-values the series converges and B_2 and B_2 coincide for barrier levels greater than $6{,}5\ \sigma_o$. The upper bounds II and III and Yang's result[30] from the Markov-assumption are very close.

It is noted that for narrowband processes and for

moderate barrier levels the Poisson assumption yields an error of about 60 % in this case. This maximal error becomes smaller for systems with higher damping values.

On the other hand, it can be observed that the Poisson process leads to nonconservative results for very low threshold levels[4]. The upper bounds are relatively high in this region.

Among all results presented in Fig.16, the upper bound II seems most appropriate for application, because it is easier to compute than Yang's approximation, and yields strictly conservative results.

3.7 Peak factors

Remembering the numerical example in Fig.13 we found for an assumed reliability those barrier levels λ for the structural response (at a specific point in space) which the structure will have to tolerate.

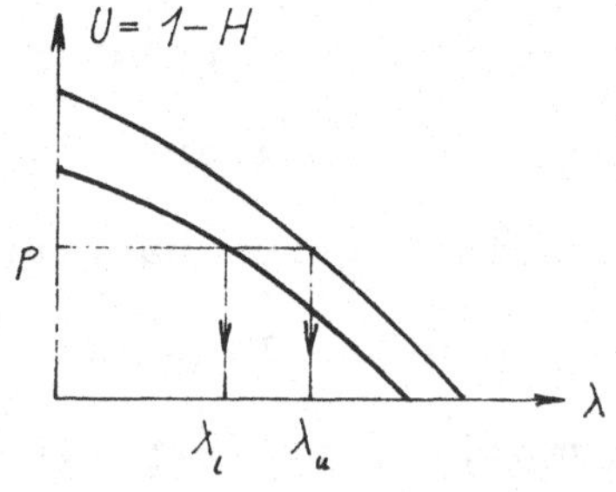

For design purpose only the upper value λ_u is interesting. λ_u can be expressed as a product of the maximum rms value of the structural response σ_1^* and the peak factor corresponding to a first-passage probability p

$$\lambda_u = \sigma_1^* \cdot r_p \tag{69}$$

Because λ_u was derived from an upper bound, it can be concluded that no more than p per cent of all realisations will exceed the threshold level λ_u and, consequently, will have an extremum greater than λ_u. Thus, the peak values relate the extreme values of the realisations with the extreme rms-value of the structural response. An approximate definition of the peak distribution of the structural response can be derived from the first-passage probability $H(\lambda)$

$$P\left[\lambda \leq \text{peak value} \leq \lambda + d\lambda\right] = \frac{dH(\lambda)}{d\lambda}$$

In what follows, the upper bound II will be used to compute the median peak factors r_{50} for one-degree-of-freedom systems with various natural frequencies and damping values under FNW excitation. The result is presented in Fig.17.

Peak factors r_{50} depend mainly on three parameters[37]:

- natural frequency ν
- fraction of critical damping ζ
- nonstationary character of the excitation $\psi = e^{-\alpha t} - e^{-\beta t}$

In the case of stationary processes the peak factors increase for a given frequency ν with increasing damping. This is due to the fact, that the bandwidth of the corresponding spectral density increases, too. In the case of nonstationary processes we also meet this tendency, but only for relatively large frequencies ν. For low and moderate frequencies we observe from Fig.18 the opposite tendency. An explanation of this fact can be given as follows. The nonstationary character of the structural response is determined from all three parameters quoted above (ν;ζ;ψ), and can be expressed by an equivalent time interval T_e defined below

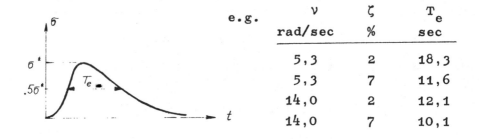

e.g.

ν rad/sec	ζ %	T_e sec
5,3	2	18,3
5,3	7	11,6
14,0	2	12,1
14,0	7	10,1

As can be seen, the equivalent time interval T_e of oscillators with low frequencies is strongly affected by the damping ratio. The higher the damping value ζ, the faster the structural response decays and, consequently, the lower the peak factors are. The difference between the peak factors corresponding to the excitations I and II can also be explained by different equivalent time intervals.

The peak factors of multi-degree-of-freedom systems

have to be computed accordingly and yield generally higher values than for single-degree-of-freedom systems.

4. FINAL REMARKS

Up to now we are able to solve the reliability problem for structures, that are modeled by linear multi-degree-of-freedom systems. Various approximate techniques have been demonstrated by means of numerical examples. If the structural response is a typical narrow-band-process, the Markov approximation for the corresponding point process equ.(63) yields the best results. For all other structures the upper bound equ.(56) should be preferred.

Last, but not least, let me state some basic remarks about nonlinear structural behaviour. Strong motion earthquakes may cause that nonlinear effects become important. Fig.19 demonstrates an example for a modern building, that was damaged during the last quake near Gemona.

We have to distinguish between nonlinear elastic behaviour (the restoring force is a unique function of deflection and velocity) and hysteretic behaviour. An excellent review of analytical methods, that are available in these cases, was recently given by Iwan[39]. In Earthquake Engineering hysteretic behaviour attracts most attention and will be briefly discussed in what follows.

Fig.18 presents the results of a simulation study of the response of elastic-ideal plastic oscillators with various yield levels, all of them excited by FNW-processes (see appendix). The first-passage probability (or the reliability) were computed from the peak distribution of the set of samples that were generated.

$$H_n(\lambda) = \frac{N(\text{peak value greater } \lambda)}{N_{total}} \tag{70}$$

The three different plots 1,2,3 belong to the yield points b_1, b_2 and b_3. The yield points are normalized to maximum rms-values σ^*_{el} of the elastic structural response. In the case of a sufficiently high yield level the structural response remains within the elastic domain (curve no.3). The graph no.2 belongs to a moderate yield level. As can be observed, the first-passage probability is less in comparison with the purely elastic case. This effect can be noticed in all those cases, where only small plastic deformations occur and is due to additional (hysteretic) damping in the structure.

If the yield point b is very low like b_1 the opposite tendency is observed. Because of accumulated plastic deformation a significantly higher first-passage probability for high barrier levels is obtained. The minor effect of additional damping leads to a reduction of $H_n(\lambda)$ only for moderate barrier levels λ.

These two contrary effects have also been observed by Caughey[40]. For practical purpose, we may therefore conclude that for relatively high yield levels ($b > 2 - 3$) conservative results are obtained, if the nonlinear hysteretic behaviour is neglected, and the structure is considered as being purely elastic.

APPENDIX

Statistics of the structural response

Herein, all needed statistics of the structural response process will be derived. Therefore, spezific models have to be selected for the excitation process $\ddot{Z}(t)$

1. Model SW: $\ddot{z} = n(t)$

$$E\{\ddot{z}(t)\ddot{z}(s)\} = J\delta(t-s)$$

the stationary excitation in modeled by Gaussian white noise.

The autocorrelation function of the j-th structural mode x_j is given from[36] as

$$R_{xx}(t,s) = \frac{Je^{-\mu_j t}}{4\nu_j}\left\{\frac{e^{\mu_j s}}{\mu_j\nu_j}\left[\cos \nu_{jd}(t-s) + \frac{\mu_j}{\nu_{jd}} \sin \nu_{jd}(t-s)\right]\right.$$

$$+ \frac{e^{-\mu_j s}}{\nu_j\nu_{jd}^2}\left[\mu_j \cos \nu_{jd}(t+s) - \nu_{jd} \sin \nu_{jd}(t+s)\right. \qquad \text{(A-1)}$$

$$\left.\left. - \frac{\nu_j}{\zeta_j} \cos \nu_{jd}(t-s)\right]\right\}$$

The variance function is

$$\sigma_x^2(t) = R_{xx}(t,t) = \frac{J}{4\mu_j\nu_j^2} + \frac{Je^{-2\mu_j t}}{4\mu_j\nu_{jd}^2\nu_j}\left[\mu_j \cos (2\nu_{jd}t) -\right.$$

$$\text{(A-2)}$$

$$\left. - \nu_{jd} \sin (2\nu_{jd}t) - \frac{\nu_j}{\zeta_j}\right] = \sigma_1^2$$

with the steady-state solution

$$\sigma_x^2 = \frac{J}{4\mu_j \nu_j^2} \tag{A-3}$$

Fig.20 shows a result[34].

The variance function of the time derivative $\dot{X}(t)$ is derived from

$$R_{\dot{x}\dot{x}}(t,t) = \frac{\partial^2 R_{xx}(t,s)}{\partial s \partial t} \Bigg|_{s=t}$$

and results in

$$\sigma_{\dot{x}}^2 = \sigma_2^2 = \frac{J}{4\mu}(1 - \frac{1+\zeta_j^2}{1-\zeta_j^2}\, e^{-2\mu_j t}) \tag{A-4}$$

2. <u>Model NW</u> $\ddot{z}(t) = n(t)\psi(t)$

$$E\{\ddot{z}(t)\ddot{z}(s)\} = J\psi^2(t)\delta(t-s)$$

the excitation is modeled by a <u>n</u>onstationary <u>w</u>hite noise (shot noise). $\psi(t)$ is a deterministic envelope function

$$\psi(t) = (e^{-\alpha t} - e^{-\beta t}) \qquad t \geq 0 \tag{A-5}$$

The autocorrelation function of the j-th structural mode x_j has to be evaluated from the integral

$$R_{xx}(t,s) = \int_0^{\min(t,s)} h_j(t-\tau) h_j(s-\tau) \psi^2(\tau) d\tau$$

and results in[4]

$$R_{xx}(t,s) = Je^{-\mu_j(s-t)} \cdot$$

$$\cdot \left[f_1(t) \cos \nu_{jd}(s-t) + f_2(t) \sin \nu_{jd}(s-t)\right] \qquad \text{(A-6)}$$

The following abbreviations were used herein

$$f_1(t) = c_{2\alpha}e^{-2\alpha t} + c_{2\beta}e^{-2\beta t} - 2c_{\alpha+\beta}e^{-(\alpha+\beta)t} +$$

$$+ e^{-2\mu_j t}\left[a_1 \cos 2\nu_{jd}t + \nu_{jd}a_2 \sin 2\nu_{jd}t + a_3\right]$$

$$f_2(t) = (d_{2\alpha}e^{-2\alpha t} + d_{2\beta}e^{-2\beta t} - 2d_{\alpha+\beta}e^{-(\alpha+\beta)t})\ \nu_{jd} \qquad \text{(A-7)}$$

$$+ e^{-2\mu_j t}\left[a_1 \sin 2\nu_{jd}t - \nu_{jd}a_2 \cos 2\nu_{jd}t\right]$$

$$\left.\begin{aligned} c_k &= \frac{2}{(2\mu_j-k)(4\nu_{jd}^2 + (2\mu_j-k)^2)} \\ d_k &= \frac{1}{\nu_{jd}^2(4\nu_{jd}^2 + (2\mu_j - k)^2)} \end{aligned}\right\} \quad k = 2\alpha,\ 2\beta,\ \alpha+\beta$$

$$a_1 = \frac{1}{2\nu_{jd}^2}\left(\frac{2\mu - 2\alpha}{4\nu_{jd}^2 + (2\mu_j-2\alpha)^2} - 2\,\frac{2\mu - \alpha - \beta}{4\nu_{jd}^2 + (2\mu_j-\alpha-\beta)^2} +\right.$$

$$\left. + \frac{2\mu - 2\beta}{4\nu_{jd}^2 + (2\mu_j-2\beta)^2}\right)$$

$$a_2 = -\,(d_{2\alpha} + d_{2\beta} - 2d_{\alpha+\beta})$$

$$a_3 = -\frac{1}{2\nu_{jd}^2}\left(\frac{1}{2\mu_j-2\alpha} + \frac{1}{2\mu_j-2\beta} - \frac{2}{2\mu_j-\alpha-\beta}\right)$$

The variance function is obtained upon putting s = t

$$\sigma_x^2(t) = J\ f_1(t) \tag{A-8}$$

and for the variance function of the time derivative $\dot{X}(t)$ and the crosscorrelation between X and $\dot{X}$ the following result is obtained

$$\sigma_{x\dot{x}}(t) = J(-\mu\ f_1(t) + \nu_{jd}\ f_2(t))$$

$$\sigma_{\dot{x}}^2(t) = J(-\mu\ f_1(t) + \nu_{jd}\ f_2(t) - (\mu_j^2+\nu_{jd}^2)f_1(t) + 2u_j\nu_{jd}f_2(t)) \tag{A-9}$$

Following Priestley[23], a wide class of processes have a spectral representation of the form

$$X(t) = \frac{1}{\sqrt{2\pi}} \int_{-\infty}^{\infty} M(t,\omega) e^{i\omega t} dG(\omega)$$

These processes possess a so-called evolutionary power spectrum and are called oscillatory processes.

The corresponding stationary process X^+ has the spectral representation

$$X^+(t) = \frac{1}{\sqrt{2\pi}} \int_{-\infty}^{\infty} e^{i\omega t} dG(\omega)$$

In our case $X^+(t)$ can be interpreted as stationary white noise and the modulating function of $\ddot{Z}(t)$ is identified with $\psi(t)$. Hence

$$\ddot{Z}(t) = \frac{1}{\sqrt{2\pi}} \int_{-\infty}^{\infty} \psi(t) e^{i\omega t} dG(\omega)$$

with

$$E\{dG(\omega_1)\ d\bar{G}\ (\omega_2)\} = J\delta(\omega_1-\omega_2)d\omega_1\ d\omega_2$$

The modulating function of $X_j(t)$ is to be computed from the truncated Fouriertransform[4]

$$M_j(t,\omega) = \int_0^t \psi(t-\tau)h(\tau)e^{-i\omega\tau}d\tau \tag{A-10}$$

and results in

$$M_j(t,\omega) = \sum_{k=1}^{2} z_{kj}\left(\frac{e^{(w_{kj}-i\omega)t} - e^{-\alpha t}}{w_{kj} + \alpha - i\omega} - \frac{e^{(w_{kj}-i\omega)t} - e^{-\beta t}}{w_{kj}+\beta-i\omega}\right) \tag{A-11}$$

with the complex numbers

$$z_{1j} = \frac{-1}{2\nu_{jd}^2} \qquad w_{1j} = -\mu_j + i\nu_{jd}$$
$$z_{2j} = \bar{z}_{1j} \qquad w_{2j} = \bar{w}_{1j} \tag{A-12}$$

being used for abbreviation.

Using this approach, the variance function of X(t) is given by

$$\sigma_x^2(t) = \frac{J}{\pi} \int_0^\infty M(t,\omega)\bar{M}(t,\omega)\,d\omega \tag{A-13}$$

and yields equ.(A-8). The integrand represents the evolutionary power spectral density and gives an impression of how the frequency content changes with time.

3. Model FNW[38] $\ddot{z}(t) = \int_0^t h_B(t-\tau)\psi(\tau)n(\tau)d\tau$ (A-14)

the excitation is modeled by filtered nonstationary white noise. $h_B(t)$ shall represent a ground filter impulse response, e.g.

$$h_B(t) = \frac{e^{-\varepsilon t}}{\omega_d} \sin \omega_d t \tag{A-15}$$

The correlation function of the input $\ddot{Z}(t)$ is identical with (A-6) if μ_j and ν_{jd} are replaced by ε and ω_d. A plot of the rms-function of $\ddot{Z}(t)$ is given in the Fig.21.

The expressions for the correlation and variance

function of X(t) are rather lengthy[4] and are not given here. A plot of one result is shown in Fig.22.

A simpler expression is obtained for the modulating function

$$M_j(t,\omega) = \sum_{k=1}^{4} z_{kj}\left(\frac{e^{(w_{kj}-i\omega)t} - e^{-\alpha t}}{w_{kj} + \alpha - i\omega} - \frac{e^{(w_{kj} - i\omega)t} - e^{-\beta t}}{w_{kj} + \beta - i\omega}\right) \tag{A-16}$$

when the following complex numbers are used

$$cz_{1j} = -2\omega_d \nu_{jd}(\mu_j-\varepsilon) - i\,\nu_{jd}((\mu_j-\varepsilon)^2+\omega_d^2+\nu_{jd}^2) + 2\nu_{jd}\omega_d^2$$

$$cz_{3j} = 2\omega_d \nu_{jd}(\mu_j+\varepsilon) - i\,\omega_d((\mu_j-\varepsilon)^2+\omega_d^2+\nu_{jd}^2) + 2\nu_{jd}^2\omega_d$$

$$c = 2\omega_d \nu_{jd}((\mu_j-\varepsilon)^2 + \omega_d^2 + \nu_{jd}^2)^2 - 4\nu_{jd}^2\omega_d^2$$

$$z_{2j} = \bar{z}_{1j}, \qquad z_{4j} = \bar{z}_{3j}$$

$$w_{1j} = -\varepsilon + i\omega_d \qquad w_{2j} = \bar{w}_{1j}$$

$$w_{3j} = -u_j + i\nu_{jd} \qquad w_{4j} = \bar{w}_{3j}$$

The autocorrelation is found from the integral

$$R_{xx}(t,s) = \frac{J}{\pi}\int_0^{\infty} M(t,\omega)\bar{M}(t,\omega)\cos\omega(s-t)\,d\omega \tag{A-17}$$

and has the following derivatives (the time and frequency dependence of the modulating function will be suppressed)

$$\sigma_1^2 k' = \left.\frac{\partial R_{xx}(t,s)}{\partial s}\right|_{s=t} = \frac{J}{\pi}\int_0^\infty \bar{M}\,\frac{\partial M}{\partial t}\,d\omega = E\left\{x(t)\dot{x}(t)\right\} \quad \text{(A-18)}$$

$$\sigma_1^2 k'' = \left.\frac{\partial^2 R_{xx}(t,s)}{\partial s^2}\right|_{s=t} =$$

$$= \frac{J}{\pi}\int_0^\infty \left(\bar{M}\,\frac{\partial^2 M}{\partial t^2} - \omega^2 M\bar{M}\right) d\omega = E\left\{x(t)\ddot{x}(t)\right\} \quad \text{(A-19)}$$

$$\sigma_1^2 s'' = \frac{\partial^2 R_{xx}(s,s)}{\partial s^2} = \frac{J}{\pi}\int_0^\infty \left(\bar{M}\,\frac{\partial^2 M}{\partial t^2} + 2\,\frac{\partial M}{\partial t}\,\frac{\partial \bar{M}}{\partial t} + M\,\frac{\partial^2 \bar{M}}{\partial t^2}\right) d\omega \quad \text{(A-20)}$$

$$\sigma_{\dot{x}}^2 = \left.\frac{\partial^2 R_{xx}(t,s)}{\partial t\,\partial s}\right|_{s=t} = \frac{J}{\pi}\int_0^\infty \left(\bar{M}M\omega^2 + \frac{\partial M}{\partial t}\,\frac{\partial \bar{M}}{\partial t}\right) d\omega \quad \text{(A-21)}$$

$$r' = \frac{J}{\pi}\int_0^\infty M\bar{M}\omega\,d\omega \quad \text{(A-22)}$$

These expressions are needed for the computation of envelope statistics.

The formulas are generally valid (e.g. for NW-model) and reduce remarkably in the stationary case, where the modulating function is identical with 1.

In order to get the statistics of the whole structural response a summation according to equ.(3) has to be performed. It the modal response variance functions σ_j^2 are known, we obtain for the total response

$$\sigma^2 = \sum_{j=1}^{N} \alpha_j^2 \sigma_j^2 + \sum_{j=1}^{N} \sum_{\substack{k=1 \\ k \neq j}}^{N} \alpha_j \alpha_k \sigma_j \sigma_k \qquad \text{(A-23)}$$

If the eigenfrequencies are well spaced, the second term can be neglected in comparison with the first one. Fig.23 shows the influence of modal interaction.

Using the modulating function equ.(A-23) can be rewritten as

$$\sigma^2 = \frac{J}{\pi} \sum_{j=1}^{N} \alpha_j^2 \int_0^{\infty} M_j(t,\omega)\, \bar{M}_j(t,\omega)\, d\omega +$$

$$+ \frac{J}{\pi} \sum_{j=1}^{N} \sum_{\substack{k=1 \\ k \neq j}}^{N} \alpha_j \alpha_k \int_0^{\infty} M_j(t,\omega)\, \bar{M}_k(t,\omega)\, d\omega$$

REFERENCES

[1] Shinozuka, M. and Sato, Y., Simulation of nonstationary random processes, Journal of the Engineering Mechanics Division, Proc.ASCE, 93, EM 1, 11, 1967.

[2] Bergstrom, R.N. and Chu, S. and Small, R.J., Seismic analysis of nuclear power plant structures, Journal of the Power Division, Proc.ASCE, 97, PO2, 367, 1970.

[3] Newmark, N.M. and Rosenblueth, E., Fundamentals of Earthquake Engineering, Prentice-Hall, Inc., 1971.

[4] Grossmayer, R., Stochastische Zuverlässigkeitsanalyse erdbebenerregter Strukturen, Doct.Thesis, Techn.Univ. Vienna, 1975.

[5] Kac, M., Probability theory as a mathematical discipline and as a tool in engineering and science, 1st Symposium on Engineering Application of Random Function Theory and Probability (Eds. J.L. Bogdanoff and F. Kozin), J.Wiley and Sons Inc., New York, 1963.

[6] Yang, J.N. and Shinozuka, M., First passage time problem, J. of Ac. Soc. of Am., 47, 1, 393, 1970.

[7] Bolotin, V.V., Statistical aspects in the theory of structural stability, Proc. Int. Conf. on Dynamic Stability of Structures (Ed.G.Herrmann), Pergamon Press, New York, 67, 1967.

[8] Ariaratnam, S.T. and Pi, H.N., On the first-passage time for envelope crossings for a linear oscillator, Int. Journal of Control, 18, 1, 89, 1973.

[9] Rosenblueth, E. and Bustamente, J.I., Distribution of the structural response to earthquakes, Journal of the Engineering Mechanics Division, Proc.ASCE, 75, 1962.

[10] Gray, Jr., A.H., First passage time problem in a random vibrational system, Journal of Applied Mechanics, 33, 167, 1966.

[11] Bogoliubov, N.N. and Mitropolsky, Y.A., Asymptotic Methods in the Theory of Nonlinear Oscillations, Gordon and Breach Science Publishers, Inc., 1961.

[12] Lennox, W.C. and Fraser D.A., On the first-passage distribution for the envelope of a nonstationary narrow-band stochastic process, Journal of Applied Mechanics, 41, 793, 1974.

[13] Crandall, S.H. and Chandiramani, K.L. and Cook, R.G., Some first passage problems in random vibration, Journal of Applied Mechanics, 33, 532, 1966.

[14] Parkus, H. and Zeman, J.L., Some stochastic problems of thermoviscoelasticity, Proc. IUTAM Symposium on Thermoinelasticity, Glasgow, 226, 1968.

[15] Bartlett, M.S., Stochastic Processes, University Press, Cambridge, 56, 1960.

[16] Roberts, J.B., An approach to the first passage problem in random vibration, Journal of Sound and Vibration, 8, 2, 301, 1968.

[17] Roberts, J.B., Probability of first passage failure for stationary random vibration, AIAA Journal, 12, 1636, 1974.

[18] Roberts, J.B., Probability of first passage failure for

nonstationary random vibration, Journal of Applied Mechanics, 42, 716, 1975.

[19] Shinozuka, M., Probability of structural failure under random loading, Journal of the Engineering Mechanics Division, Proc. ASCE, 90, EM 5, 147, 1964.

[20] Lin, Y.K., Probabilistic Theory of Structural Dynamics, McGraw-Hill, New York, 1967.

[21] Cramer, H. and Leadbetter, M.R., Stationary and Related Stochastic Processes, Wiley, New York, 1967.

[22] Rice, J.R. and Beer, F.P., First occurence time of high-level crossings in a continuous random process, Journal of Ac. Soc. of Am., 39, 2, 323, 1966.

[23] Priestley, M.B., Power spectral analysis of nonstationary random processes, Journal of Sound and Vibration, 6, 86, 1967.

[24] Rice, S.O., Mathematical analysis of random noise, in Selected Papers on Noise and Stochastic Processes (Ed. N.Wax), New York-Dover, 1955.

[25] Yang, J.N., Simulation of random envelope process, Journal of Sound and Vibration, 21, 1, 73, 1972.

[26] Abramowitz, M. and Stegun, I.A., Handbook of Mathematical Functions, Dover Publications, Inc., New York, 1949.

[27] Vanmarcke, E.H., On the distribution of the first-passage time for normal stationary random processes, Journal of Applied Mechanics, 42, 215, 1975.

[28] Grossmayer, R., On the application of various crossing probabilities in the aseismic reliability problem, IUTAM Symposium on Stochastic Problems in Dynamics, Southampton, 1976.

[29] Yang, J.N. and Shinozuka, M., On the first excursion probability in stationary narrow-band vibration, Journal of Applied Mechanics, part I, 38, 1017; part II, 39, 733, 1972.

[30] Yang, J.N., First-excursion probability in nonstationary random vibration, Journal of Sound and Vibration, 27, 2, 165, 1973.

[31] Lin, Y.K., First excursion failure of randomly excited structures, AIAA Journal, 8, 4, 720, 1970; 8, 10, 1888, 1970.

[32] Wiener, N., Generalised Harmonic Analysis, Acta Mathematica, Stockholm, 55, 117.

[33] Lyon, R.H., On the vibration statistics of a randomly excited hard-spring oscillator, Journal of the Ac. Soc. of Am., 33, 1395, 1961.

[34] Caughey, T.K. and Stumpf, H.J., Transient response of a dynamic system under random excitation, Journal of Applied Mechanics, 28, 563, 1961.

[35] Barnoski, R.L. and Maurer, J.R., Mean-square response of simple mechanical systems to nonstationary random excitation, Journal of Applied Mechanics, 36, 221, 1969.

[36] Parkus, H., Random processes in mechanical sciences, Springer-Verlag, Wien-New York, CISM Courses and Lectures, 9, 1969.

[37] Grossmayer, R., Spitzenwertantwortspektren für erdbebenerregte Schwinger, ZAMM, T-33, 1976.

[38] Levy, R. and Kozin, F. and Moorman, R.B.B., Random processes for earthquake simulation, Journal of the Engineering Mechanics Division, Proc. ASCE, 97, 2, 495, 1971.

[39] Iwan, W.D., Application of nonlinear analysis techniques, in Applied Mechanics in Earthquake Engineering, AMD 8, Proc. ASCE, (Ed. W.D. Iwan), 135, 1974.

[40] Caughey, T.K., Random excitation of a system with bilinear hysteresis, Journal of Applied Mechanics, 27, 649, 1960.

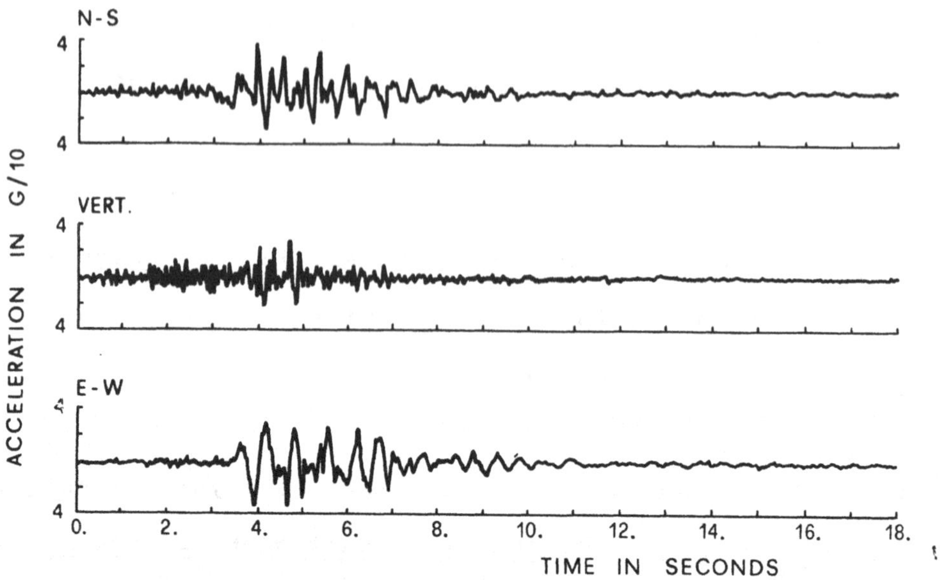

Fig.1 Earthquake Records from Tolmezzo, Italy, 6th May 1976.

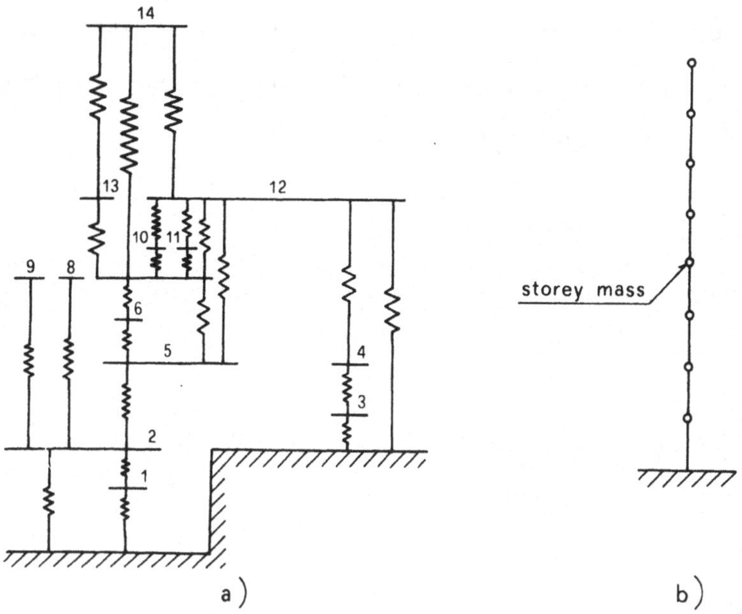

Fig.2 Lumped-Parameter Models of a Nuclear Power Plant (a) and a Multistorey-Building (b).

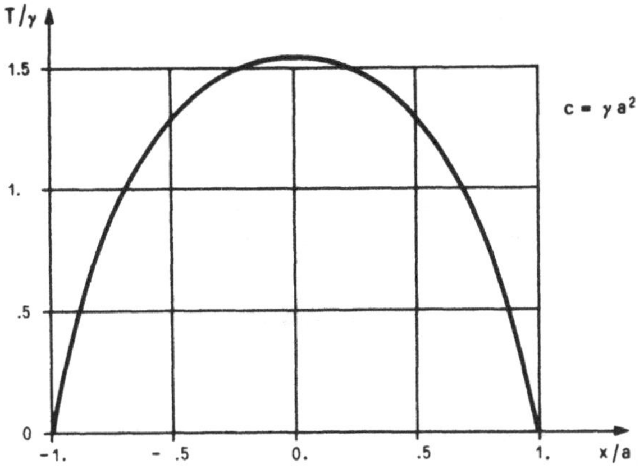

Fig.3 Mean First-Passage Time T versus Starting Position x.

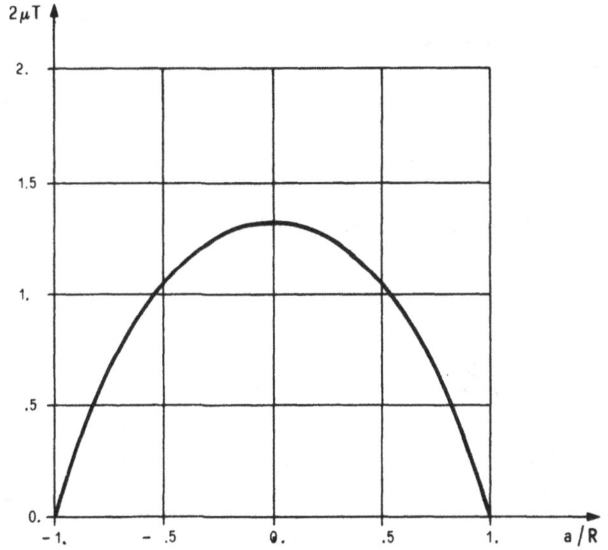

Fig.4 Mean First-Passage Time T for the Envelope Process versus Starting Position a.

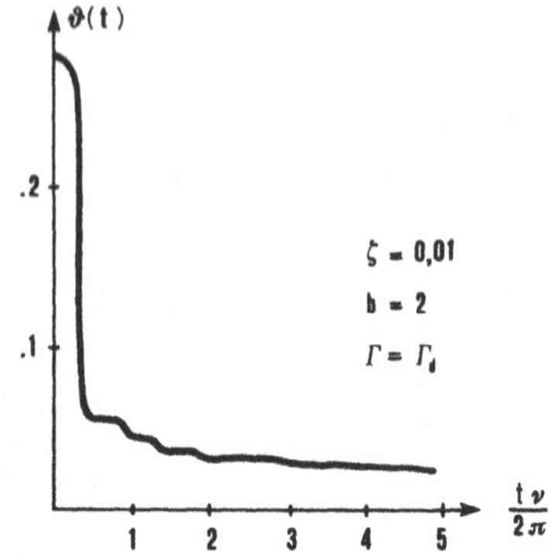

Fig.5 First-Passage Probability Density ϑ for Stationary Start [13].

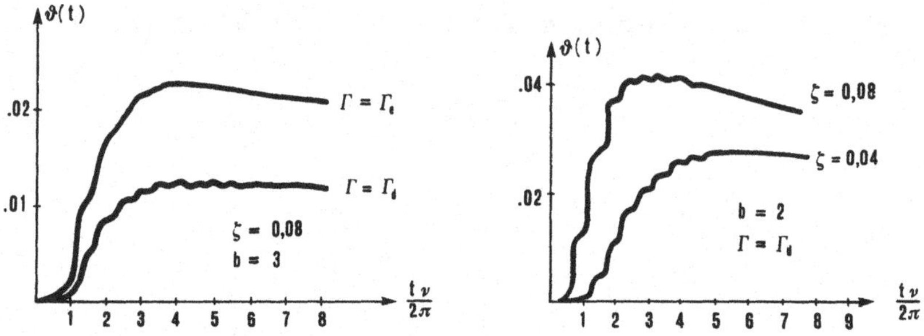

Fig.6 First-Passage Probability Density ϑ for Zero Start [13].

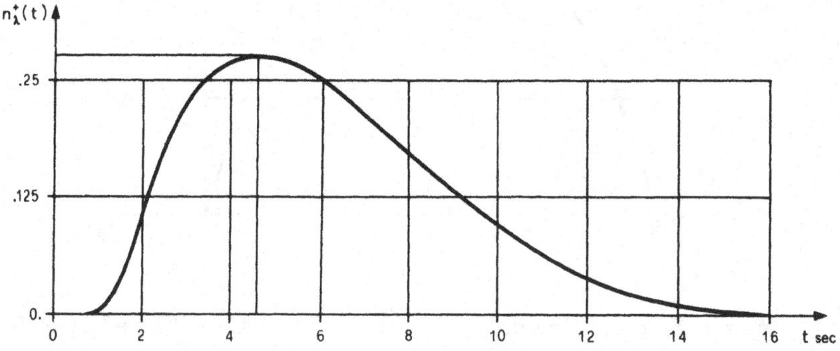

Fig.7 Time-Behaviour of the Mean Rate of Upward Crossings of the Barrier Level $\lambda = 1{,}5\ \sigma^*$ for the Structural Response Process $x(t)$; $\nu = .84$ Hz, $\zeta = 5$ %, FNW-Excitation: $\alpha = .1$, $\beta = 1.$, $\omega_d = 18.1$, $\varepsilon = 10.$, $J = 500$.

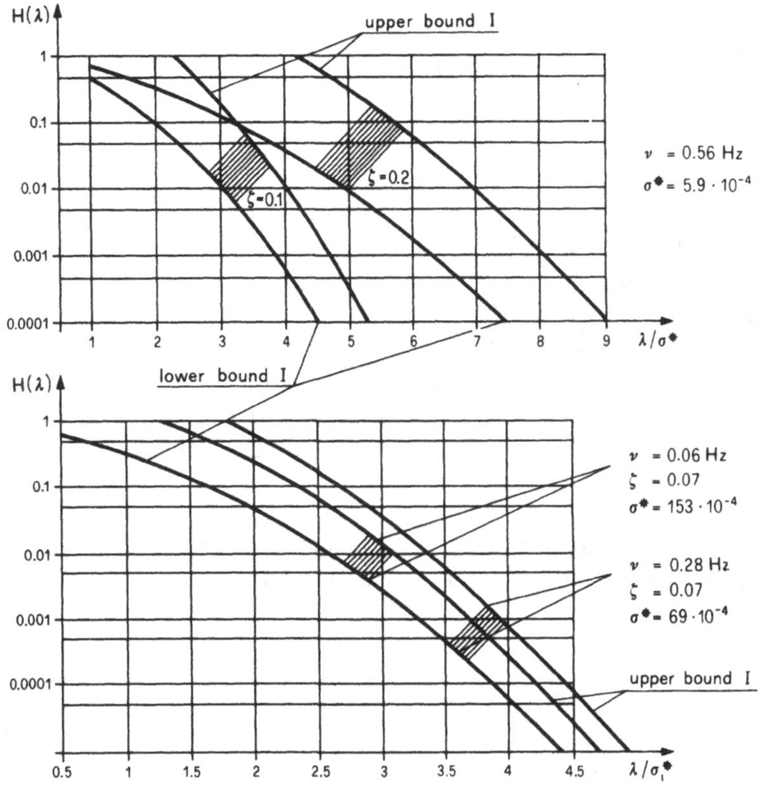

Fig.8 Upper and Lower Bounds I for the First-Passage Probability $H(t \rightarrow \infty;\ \lambda)$ for Various Linear Damped Oscillators.

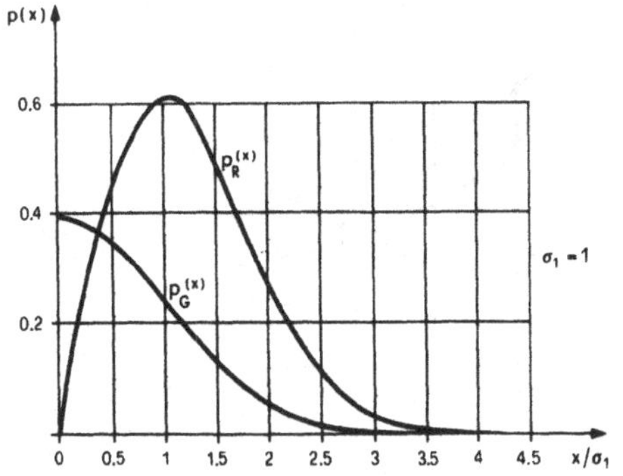

Fig.9 Rayleigh and Gaussian Probability Distribution.

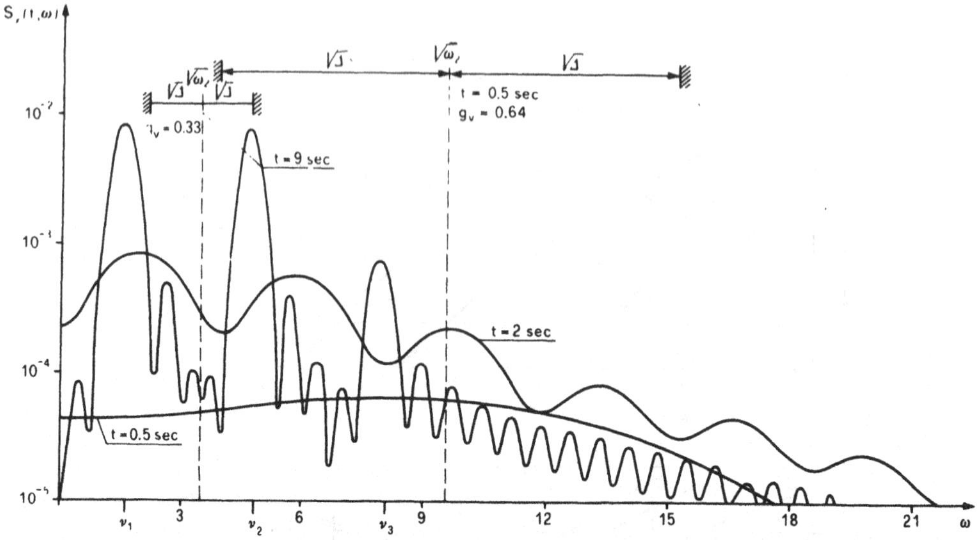

Fig.10 Evolutionary Power Spectral Density $S_{\dot{x}}(t,\omega)$ for a MDOF System (T = 4., 1.33, .8 sec., ζ = 5 %).

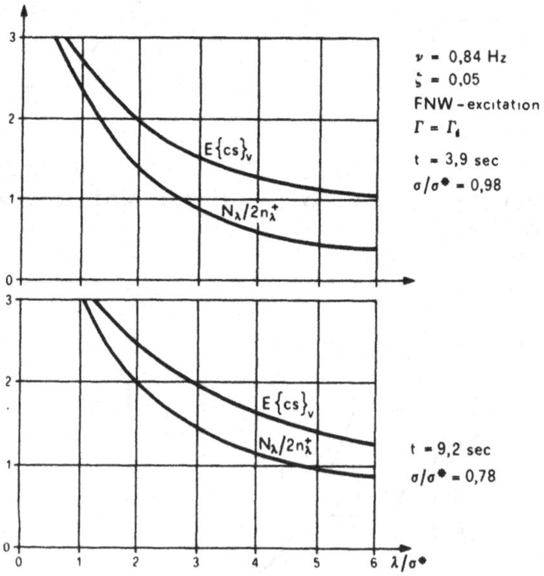

Fig.11 Mean Clumpsize for a Linear Damped Oscillator at Two Different Time Instants.

$\nu_1 = 0{,}84$ Hz $\zeta_1 = 0{,}02$

$\nu_2 = 5.26$ Hz $\zeta_2 = 0{,}003$

$\nu_3 = 14.72$ Hz $\zeta_3 = 0{,}001$

$X(t) = w(\xi = 1)$ $\sigma_w^* = .0125\,l$

$X(t) = Q(\xi = 0)$ $\sigma_a^* = .605\,G$ base shear

$X(t) = M(\xi = 0)$ $\sigma_M^* = .331\,Gl$ base moment

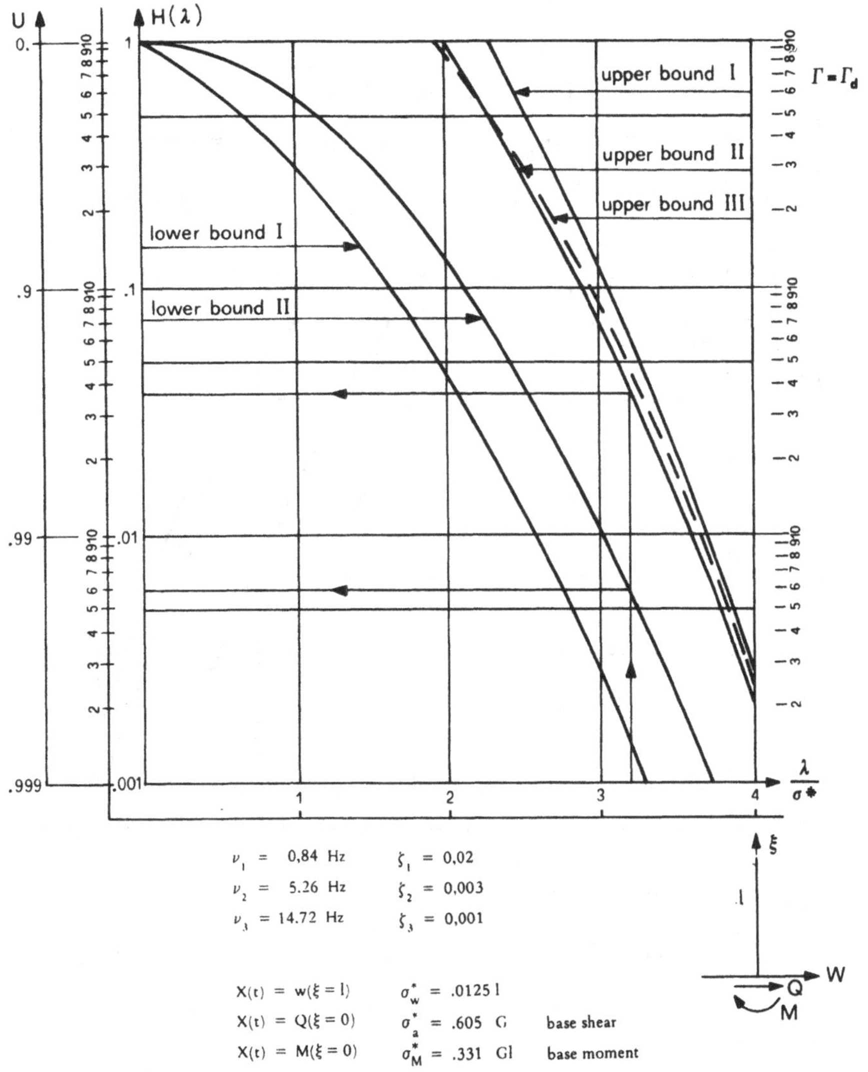

Fig. 12   Upper and Lower Bounds for the First-Passage
Probability $H(t \longrightarrow \infty, \lambda)$ for a Flexural Beam Model ($c = \sqrt{EJ/\rho F} = 1.5$) under FNW-Excitation:
$\alpha = .1$, $\beta = 1.$, $\varepsilon = 10.$, $\omega_d = 18.1$, $\sigma_{\ddot{z}}^* = .12$ g.

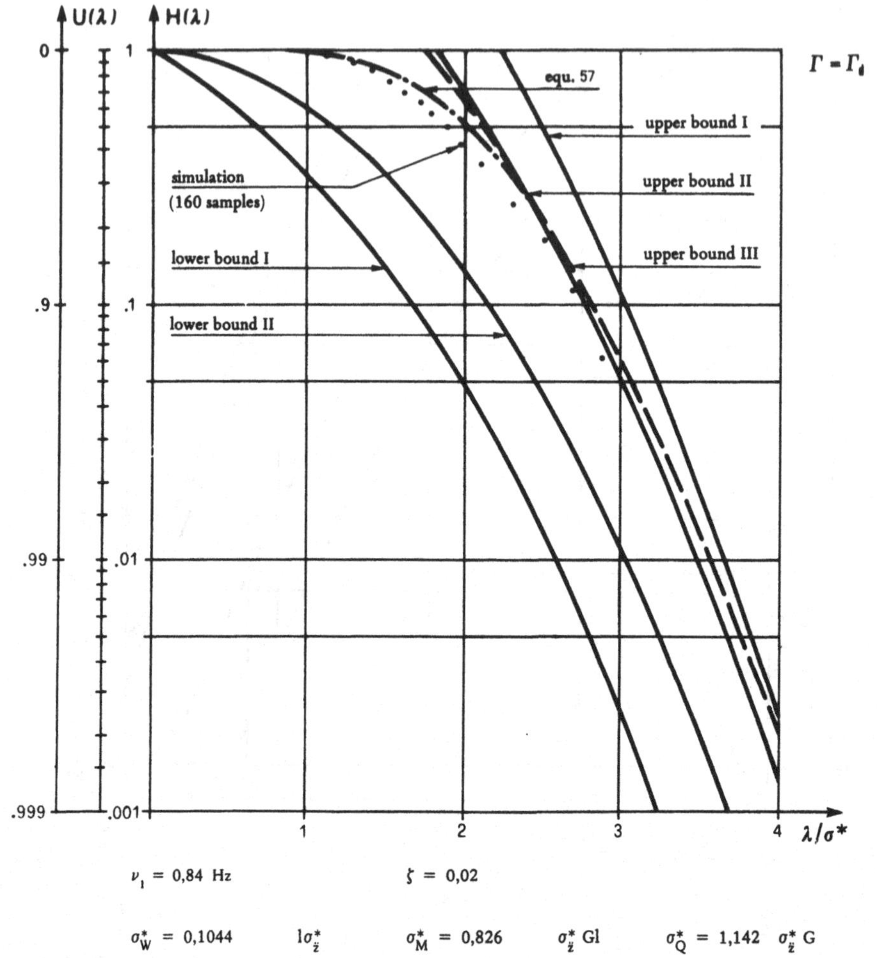

Fig.13 Upper and Lower Bounds for the First-Passage Probability $H(t \to \infty; \lambda)$ for a SDOF System identical to the First Mode of Fig.12

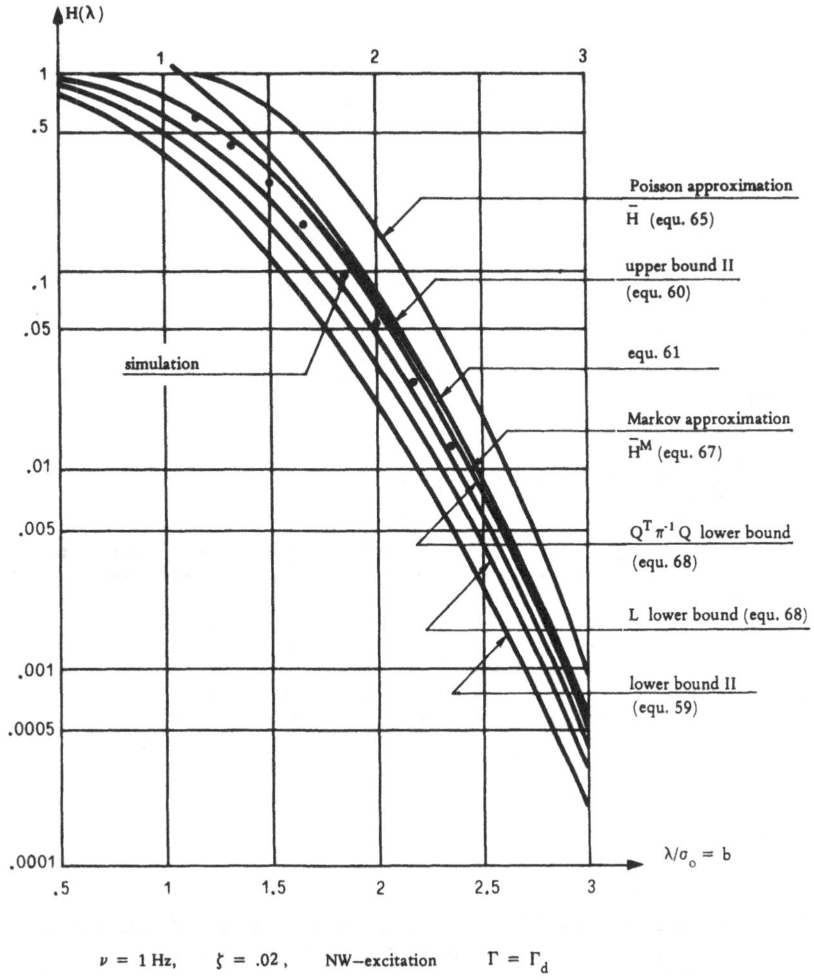

Fig. 14 Variation of the First-Passage Probability $H(t \to \infty;\ \lambda)$ obtained from Point Approximations and from Upper and Lower Bound II under NW-Excitation.

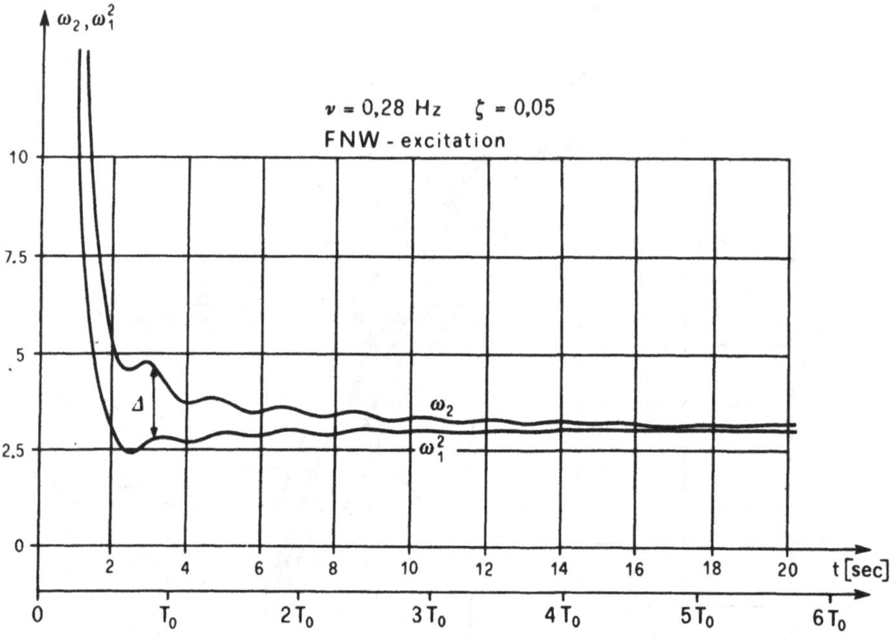

Fig.15 Variation of Midfrequency ω_2 and of ω_1^2 and, consequently of the Bandwidth with Time.

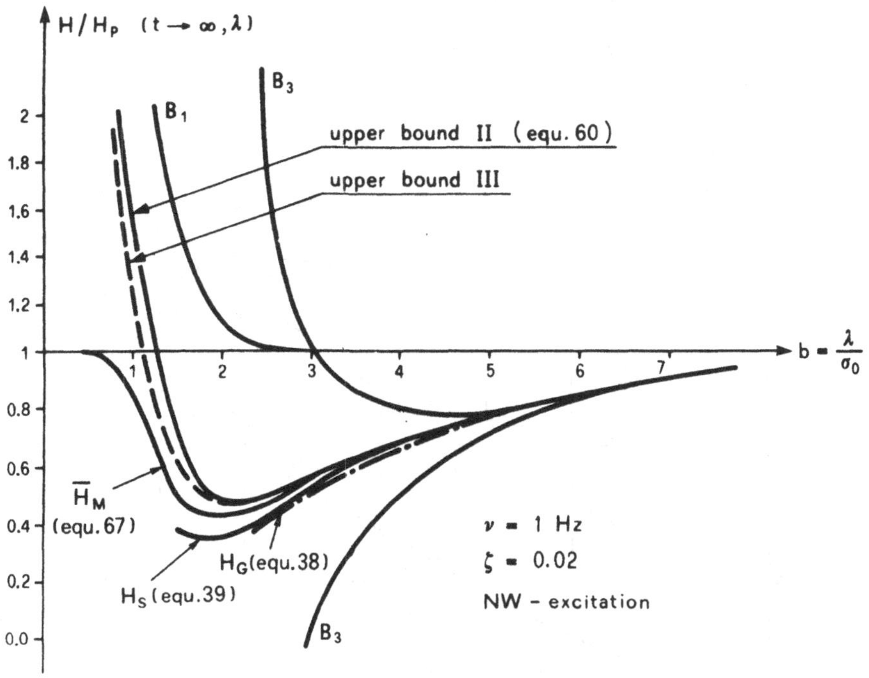

Fig. 16 Various Results for the First-Passage Probability $H(t \to \infty; b)$ normalized to the Poisson Approximation H_p(B_i series equ. 34 truncated after the i-th term).

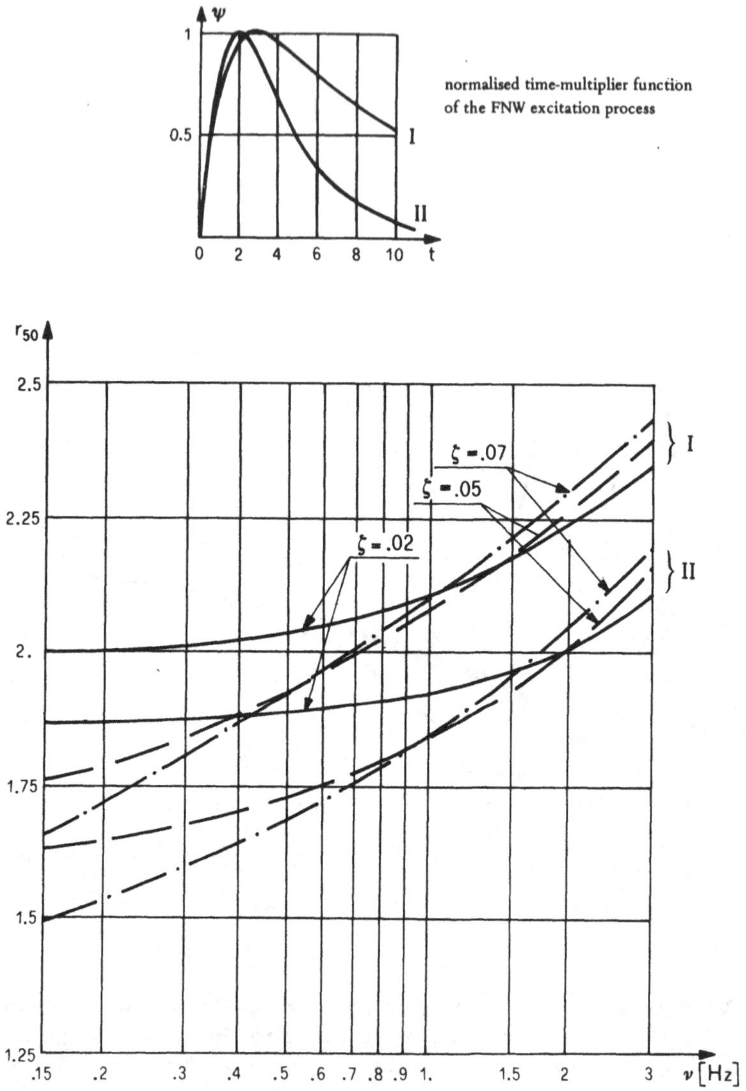

Fig.17 Median Peak Factor Spectra r_{50} for Two Different FNW-Excitations

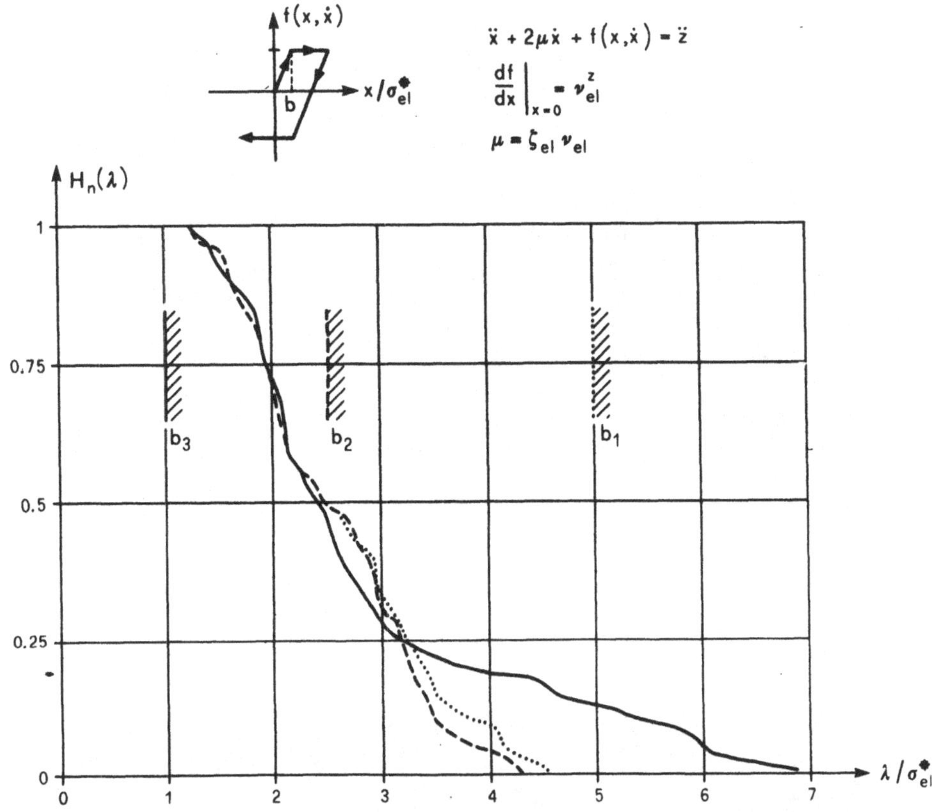

Fig.18 Simulation Results for the First-Passage Probability $H_n(\lambda;b)$ of Elastic-Plastic Oscillators under FNW-Excitation (ν_{el} =.5 Hz, ζ_{el} = 2 %, N_{total}= 160).

Fig.19 Modern Building near Gemona, damaged during the Friuli Earthquake, 6th May 1976.

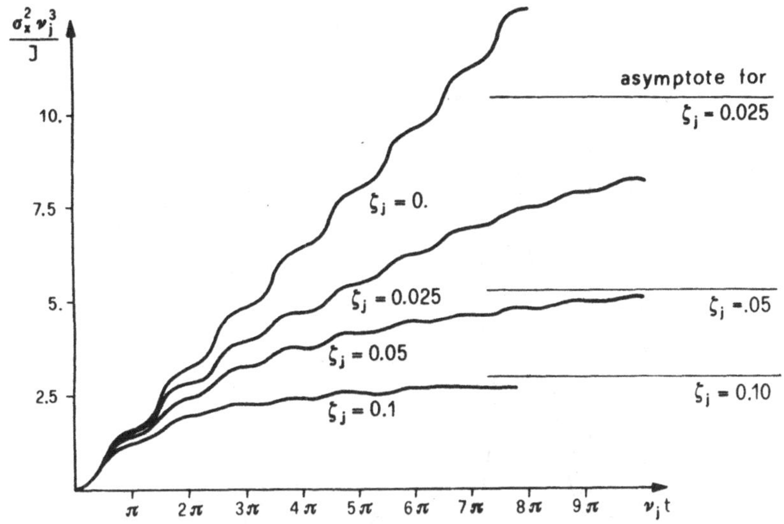

Fig.20 Normalized Mean-Square Response for Linear, Damped Oscillators under SW-Excitation

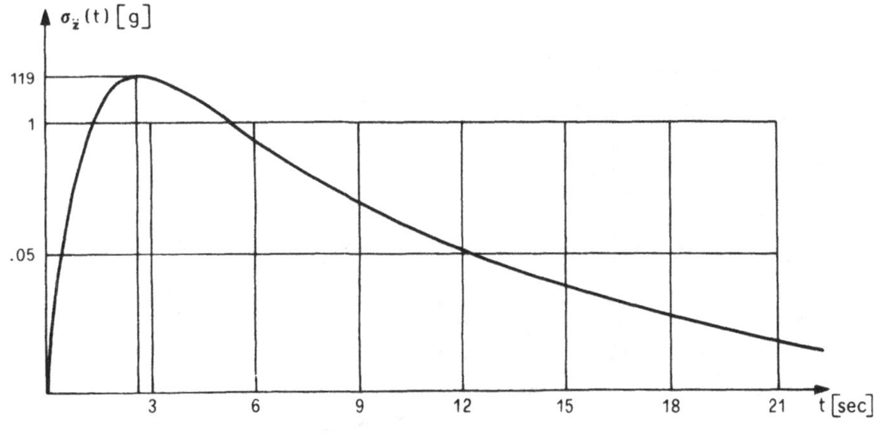

Fig.21 Rms-Ground Acceleration $\sigma_{\ddot{z}}(t)$ for FNW-Excitation ($J = 500.$, $\varepsilon = 10.$, $\omega_d = 18.1$, $\alpha = .1$, $\beta = 1.$).

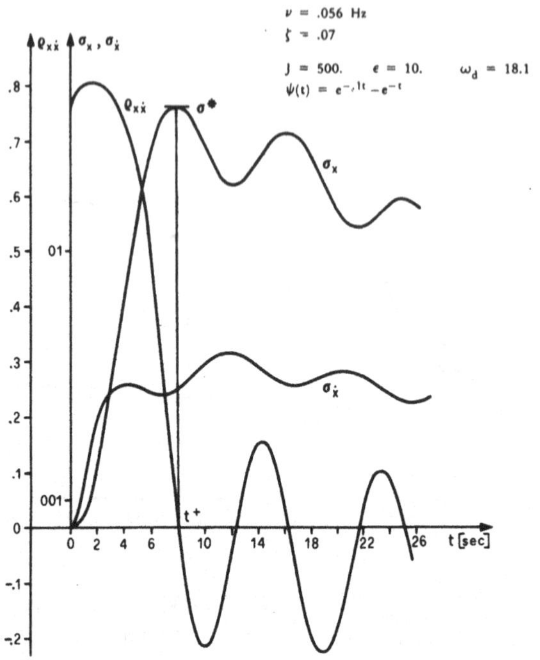

Fig.22 Rms-Structural Response of x(t), ẋ(t) and their Correlation Coefficient

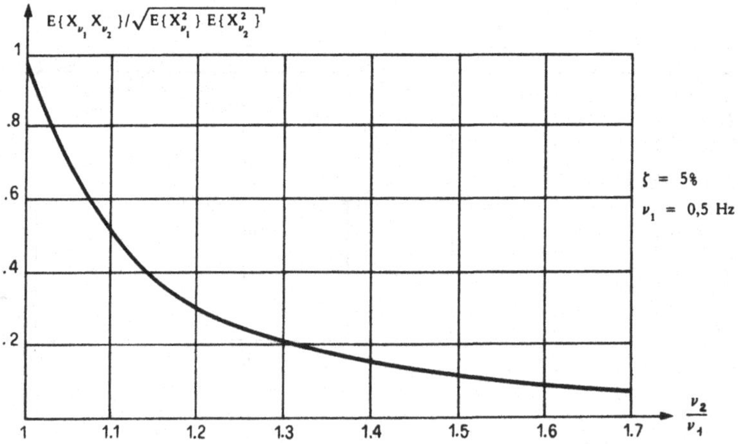

Fig.23 Correlation Coefficient of the Structural Responses $x_\nu(t)$ corresponding to Linear Damped Oscillators with an Undamped Natural Frequency ν_j under NW-Excitation

APPLICATIONS OF DIGITAL SIMULATION OF GAUSSIAN RANDOM PROCESSES

MASANOBU SHINOZUKA
Columbia University
New York, N. Y., U. S. A.

1. INTRODUCTION

In the last two decades, much research effort has been devoted to the application of the stochastic process theory in the general area of engineering mechanics and structural engineering for the purpose of predicting the dynamic structural performance with a better accuracy and of assessing the over-all structural safety with a better reliability by considering more realistic analytical models of load-structure systems.

Naming only a few, possible applications of this stochastic approach include (a) analysis of panel vibrations of aircraft induced by boundary layer turbulence, (b) analysis of ship motions caused by ocean waves, particularly during a storm, (c) analysis of aircraft response to gust vertical velocity, (d) response analysis of off-shore structures to wave and wind forces,(e) statistical strength analysis of engineering materials with randomly distributed material properties, (f) analysis of the effect of the randomness in geometrical configuration and/or in mechanical constraint of a structural component (due, for example, to fabrication errors) on the vibration and buckling eigenvalues, (g) study of random surface roughness of bridge pavement and airport run-way for the purposes of analyzing the vehicle and aircraft vibration caused by the roughness and of stress analysis of pavement systems under the action of vehicles and aircraft, and of course, (h) response analysis of structures subject to earthquake acceleration or to atmospheric turbulence.

In spite of the recent remarkable advance in these areas of study, however, the present state of the art still leaves a number of difficulties unsolved that must be overcome before the approach becomes more useful. Such problem areas include

(1) random response analysis of highly nonlinear structures,

(2) failure analysis of structures under random loading,

(3) analysis of extreme values and,

(4) property problems associated with random material.

The recent advent of high speed digital computers, however, has made it not only possible but also highly practical to apply Monte Carlo techniques to a large variety of engineering problems including those mentioned above. This paper presents a technique of digital simulation of Gaussian random processes and associated envelope processes. Such simulations are accomplished by summing a large number of cosine functions with weighted amplitudes and random phase angles and used as the basic tool in a general Monte Carlo method of solution to a wide class of practical problems in engineering, particularly those mentioned above. See, for example, References 1-10.

2. SIMULATION OF A RANDOM PROCESS

A basic representation of a homogeneous Gaussian (one-dimensional and one-variate) random process $f_o(x)$ with mean zero and spectral density $S_o(\omega)$ in the form of the sum of the cosine functions has existed for some time[22].

$$f(x) = \sqrt{2} \sum_{k=1}^{N} A_k \cos (\omega_k x - \Phi_k) \tag{1}$$

where Φ_k are random angles distributed uniformly between 0 and 2π and independent of $\Phi_j (k \neq j)$, and

$$A_k = \left[S_1^o(\omega_k)\Delta\omega\right]^{1/2}, \qquad \omega_k = k\Delta\omega \tag{2}$$

with

$$S_1^o(\omega) = 2S_o(\omega) \tag{3}$$

being the one-sided spectral density function (see Fig.1). For digital simulation of a sample function $\bar{f}(x)$ of $f(x)$ and therefore of $f_o(x)$, Eq.1 is used with Φ_k being replaced by their realized values φ_k;

$$\bar{f}(x) = \sqrt{2} \sum_{k=1}^{N} A_k \cos (\omega_k x - \varphi_k) \tag{4}$$

with the aid of Eqs.1 and 4 it is easy to show that Eq.1 is ergodic at least up to the second moment.

Speaking of structure-related applications, however, until Borgman[23] published a paper simulating ocean surface elevation as a multidimensional process essentially in the same form, little attention had been paid to this representation in spite of its substantial advantage over the standard method in which a random process was digitally generated as output of an appropriate (analytical) filter subjected to a simulated white process. The use of this filtering technique, although limited in its practical applications to a one-variate one-dimensional process has dominated a large number of papers involving simulations of a random process.

Practical digital simulation of a multidimensional process has been made possible, as mentioned above, by Borgman[23] in principle through the preceding expression consisting of a sum of cosines and by Shinozuka[1] whose method reduces, in case of one-variate one-dimensional situations to the use of the following expression;

$$f(x) = \sigma \left(\frac{2}{N}\right)^{1/2} \sum_{k=1}^{N} \cos (\omega_k x - \varphi_k) \tag{5}$$

where φ_k are as previously defined and ω_k are realized values of random frequencies distributed according to the density function $g(\omega) = S_o(\omega)/\sigma^2$ with

$$\sigma^2 = \int_{-\infty}^{\infty} S_o(\omega) d\omega \tag{6}$$

Borgman[23] and Shinozuka[1,2,11] also investigated the digital simulation of the multivariate (but one-dimensional) process, the former making use of the filtering technique and the latter in the more convenient form of the sum of the cosine functions.

Reference 2 shows that the autocorrelation function $R(\xi)$ of $f(x)$ given in Eq 1 converges to $R_o(\xi)$ of $f_o(x)$ in the form of $1/N^2$ as $N\to\infty$. The same trend in convergence can be shown[2] to exist between the spectral element $S(\omega)\Delta\omega$ of $f(x)$ given in Eq.1 and $S_o(\omega)$ of $f_o(x)$, whereas the spectral element of $f(x)$ given in Eq.5 converges much more slowly to the target spectral element[1].

There is a constraint to be imposed upon $\Delta\omega$ in Eq.2. This constraint is due to the periodicity of the sample function $\bar{f}(x)$. Obviously, the period of $\bar{f}(x)$ is $T_o = 2\pi/\Delta\omega$ and therefore, depending on the purpose of the simulation, $\Delta\omega$ has to be so chosen that T_o is long enough for that purpose.

A significant improvement in the efficiency of digital simulation has been suggested by Yang[3] in the following form:

$$\bar{f}(x) = \sqrt{\Delta\omega}\ \mathrm{Re}\ F(x) \tag{7}$$

in which Re F(x) represents the real part of F(x) and

$$F(x) = \sum_{k=0}^{N-1}\left\{\left[2\ S_1^o(\omega_k)\right]^{1/2} e^{-i\varphi_k}\right\} e^{i\omega_k x} \tag{8}$$

is the finite complex Fourier transform of $\left[2\ S_1^o(\omega_k)\right]^{1/2} e^{-i\varphi_k}$ with ω_k and φ_k defined in Eqs.1 and 2. The advantage of Eq.8 is such that the function F(x) can readily be computed by applying the Fast Fourier Transform (FFT) algorithm, hence avoiding the time-consuming computation of a large number of cosine functions.

In the preceding discussion, the spacing $\Delta\omega$ in the frequency domain has been taken as constant. This, however, does not necessarily have to be always observed. It is possible and in fact may even be advisable to use variable spacings depending on the extent of fluctuation of the spectral density to optimize the number of cosine terms in the summation in Eq.1 with finer spacing in those domains of frequency where the fluctuation of the spectral density is more rapid and coarser spacings elsewhere. It is also likely that such variable spacings will increase the period T_o of the simulated process. If this is done, however, the simulation through the FFT technique does not appear to be possible.

In the same reference 3, Yang also proposed to simulate the envelope process $v_o(x)$ of a random process $f_o(x)$ by following the definition of the envelope process[24].

$$v_o(x) = \left[f_o^2(x) + \hat{f}_o^2(x)\right]^{1/2} \tag{9}$$

where $\hat{f}_o(x)$ is the Hilbert transform of $f_o(x)$ and, in the present case, can be written as

$$\hat{f}_o(x) = \int_0^\infty \left[\sin \omega x \, dU(\omega) - \cos \omega x \, dV(\omega)\right] \tag{10}$$

where $U(\omega)$ and $V(\omega)$ are mutually orthogonal, real processes with orthogonal increments.

It then follows that $\hat{f}_o(x)$ can be simulated as $\hat{f}(x)$:

$$\hat{f}(x) = \sqrt{2} \sum_{k=1}^{N} A_k \sin(\omega_k x - \Phi_k) \tag{11}$$

Hence, the envelope process $v_o(x)$ can be simulated as $v(x)$:

$$v(x) = \left[f^2(x) + \hat{f}^2(x)\right]^{1/2} \tag{12}$$

As an example, consider the response process $y_o(t)$ of a single degree of freedom system to a Gaussian white noise excitation $x_o(t)$ with constant spectral density S_o;

$$\ddot{y}_o(t) + 2\zeta\omega_o\dot{y}_o(t) + \omega_o^2\, y_o(t) = x_o(t) \tag{13}$$

It is well known that the spectral density of $y_o(t)$ is

$$S_y(\omega) = \frac{S_o}{(\omega^2 - \omega_o^2)^2 + 4\zeta^2\,\omega_o^2\,\omega^2} \tag{14}$$

and the standard deviation σ_y of $y_o(t)$ is

$$\sigma_y = \left(\frac{\pi S_o}{2\zeta\,\omega_o^3}\right)^{1/2} \tag{15}$$

With the aid of Eqs.1, 11, and 12, a segment of sample function $\bar{y}(t)$ of simulated process $y(t)$ for $y_o(t)$ and that of simulation $v(t)$ of the envelope process $v_o(t)$ are computed and shown in Fig.2 and 3. Figure 2 is for the case where the damping coefficient $\zeta = 0.02$ and hence the process $y_o(t)$

is narrow-band. This fact is well demonstrated by the smooth behavior of the sample envelope. When the local maxima of the sample envelope do not coincide with the local maxima (peaks) of the process, they reflect the local minima (troughs) of the simulated process. Fig.3 shows the sample functions of v(t) and y(t) for $\zeta = 0.5$. In this case, the process $y_o(t)$ is substantially wide-band and this fact is clearly seen from the much wilder fluctuation of both simulated envelope and simulated process, although the simulated envelope surprisingly well reflects peaks and troughs of the simulated process even though the process $y_o(t)$ is wide-band. In terms of simulation efficiency, the computer time will be significantly reduced if one is interested in peak- and trough - values of the process and if the process is narrow-band, since then the smooth nature of the envelope function makes it possible to use much larger interval between successive time instants at which the values of the simulated process is evaluated. The order of magnitude of this interval can be that of the apparent period of the process, which obviously is much too large for simulation of the process itself.

Again, following Yang[3], $\hat{f}(t)$ is written as

$$\hat{f}(t) = \sqrt{\Delta\omega}\ \mathrm{Im}\ F(t) \tag{16}$$

and hence v(t) can also be simulated through the FFT technique.

The Gaussian property of the simulated process (Eq.1) comes from the central limit theorem because it consists of a sum of a large number of independent functions of time. Obviously, the larger the value of N, the closer the simulated process approaches Gaussian. Recent paper by Yang[25] indicates that if, for example, N = 500 is used in Eq.1, the simulated process differs approximately 1 %, 2 % and 5 % from the Gaussian process in terms of the first-order probability distribution at 3σ, 4σ and 5σ level, respectively. Since the use of FFT makes it not only practical but also routine to use the value of N as large as 1,000 or more, the method of simulation considered here is the only realistic method of evaluating some of the hard-to-get quantities (e.g. first excursion probability).

3. SIMULATION OF A MULTIDIMENSIONAL HOMOGENEOUS PROCESS

The autocorrelation function of an n-dimensional homogeneous real process $f_o(\underline{x})$ defined by

$$R_o(\underline{\xi}) = E\left[f_o(\underline{x}_1)\ f_o(\underline{x}_2)\right]$$

is even in $\underline{\xi}$ (symmetric with respect to the origin of the n-dimensional space).

$$R_o(\underline{\xi}) = R_o(-\underline{\xi}) \tag{17}$$

where $\underline{x}_1$ and $\underline{x}_2$ are space vectors and $\underline{\xi} = \underline{x}_2 - \underline{x}_1$ is the separation vector. Assume that the n-fold Fourier transform of $R_o(\underline{\xi})$ exists. The spectral density function of $f_o(\underline{x})$ is then defined as

$$S_o(\underline{\omega}) = \frac{1}{(2\pi)^n} \int_{-\infty}^{\infty} R_o(\underline{\xi}) e^{-i\underline{\omega} \cdot \underline{\xi}} \, d\underline{\xi} \tag{18}$$

where $\underline{\omega}$ is the frequency (wave number) vector and $\underline{\omega} \cdot \underline{\xi}$ is the inner product of $\underline{\omega}$ and $\underline{\xi}$, and for simplicity

$$\int_{-\infty}^{\infty} (\ ) d\xi \equiv \int_{-\infty}^{\infty} \overset{\text{n-fold}}{\cdots\cdots\cdots\cdots} \int_{-\infty}^{\infty} (\ ) d\xi_1 \, d\xi_2 \ldots d\xi_n$$

with n being the dimension of the vector $\underline{\xi}$. It follows from Eq.17 that

$$\int_{-\infty}^{\infty} R_o(\underline{\xi}) \sin (\underline{\omega} \cdot \underline{\xi}) \, d\underline{\xi} = 0$$

and, therefore, from Eq.18

$$S_o(\underline{\omega}) = S_o(-\underline{\omega}) \tag{19}$$

Then

$$S_o(\underline{\omega}) = \frac{1}{(2\pi)^n} \int_{-\infty}^{\infty} R_o(\underline{\xi}) \cos (\underline{\omega} \cdot \underline{\xi}) \, d\xi \tag{20}$$

and is real.

It can be shown[26] that $R_o(\underline{\xi})$ is non-negative definite and therefore it has a non-negative n-fold Fourier transform;

$$S_o(\omega) \geq 0 \tag{21}$$

Based on these properties of $S_o(\underline{\omega})$, a method of simulating $f_o(\underline{x})$ is proposed in the following:

Consider an n-dimensional homogeneous process with mean zero and spectral density function $S_o(\underline{\omega})$ which is of insignificant magnitude outside the region defined by

$$-\infty < \underline{\omega}_\ell \leq \underline{\omega} \leq \underline{\omega}_u < \infty$$

and denote the interval vector by

$$(\Delta\omega_1, \Delta\omega_2, \ldots, \Delta\omega_n) = \left(\frac{\omega_{1u}-\omega_{1\ell}}{N_1}, \frac{\omega_{2u}-\omega_{2\ell}}{N_2}, \frac{\omega_{nu}-\omega_{n\ell}}{N_n}\right) \tag{22}$$

where usually $\underline{\omega}_\ell = -\underline{\omega}_u$. Then the process can be simulated by the series

$$f(\underline{x}) = \sqrt{2} \sum_{k_1=1}^{N_1} \sum_{k_2=1}^{N_2} \sum_{k_n=1}^{N_n} \left[S_o(\omega_{1k_1}, \omega_{2k_2}, \ldots, \omega_{nk_n})\Delta\omega_1\Delta\omega_2\ldots\Delta\omega_n\right]^{\frac{1}{2}}$$

$$\cdot \cos(\omega_{1k_1}x_1 + \omega_{2k_2}x_2 + \ldots \omega_{nk_n}x_n + \Phi_{k_1k_2\ldots k_n}) \tag{23}$$

where

$\Phi_{k_1k_2\ldots k_n}$ = independent random phase uniformly distributed between 0 and 2π

$$\omega_{ik_i} = \omega_{i\ell} + k_i \Delta\omega_i \qquad k_i = 1,2,\ldots N_i \qquad i = 1,2,\ldots,n$$

As in the one-dimensional case, the digital simulation $\bar{f}(\underline{x})$ of $f(\underline{x})$ can be achieved by using Eq.23 with $\Phi_{k_1k_2\ldots k_n}$ being replaced by their realized values $\varphi_{k_1k_2\ldots k_n}$.

To avoid the lengthy expressions in the subsequent discussion, $\bar{f}(\underline{x})$ will be written in the following compact form:

$$\bar{f}(\underline{x}) = \sqrt{2} \sum_{k=1}^{N} A(\underline{\omega}_k)\cos\,(\underline{\omega}_k \cdot \underline{x} + \varphi_k) \tag{24}$$

where

$$N = N_1N_2\ldots N_n$$

$$A(\underline{\omega}_k) = \left[S_o(\underline{\omega}_k)\Delta\omega_1\Delta\omega_2\ldots.\Delta\omega_n\right]^{1/2} = \left[S_o(\underline{\omega}_k)\Delta\underline{\omega}\right]^{1/2} \tag{25}$$

It is noted that if the symmetric condition of $S_o(\underline{\omega})$ is used, N in Eq.23 can be reduced by one-half. Furthermore, if the process is isotropic, N is reduced to $\frac{N}{2^n}$. Fig.4 illustrates the significance of $A(\underline{\omega}_k)$ for two-dimensional cases where, however, $A_{k_1k_2}$ is written for $A(\omega_{1k_2}, \omega_{2k_2})$.

It can be shown[2] that the ensemble average of $f(x)$ is zero, and the autocorrelation function $R(\underline{\xi})$ of $f(\underline{x})$, becomes

$$R(\underline{\xi}) = \sum_{k=1}^{N} A^2(\underline{\omega}_k)\cos\,(\underline{\omega}_k \cdot \underline{\xi}) \tag{26}$$

Upon substituting $A^2(\underline{\omega}_k) = S_o(\underline{\omega}_k)\Delta\omega$, and taking the limit as $N \to \infty$ (in the sense that $N_1, N_2, \ldots, N_n \to \infty$ simultaneously), one obtains

$$R(\underline{\xi}) = \int_{-\infty}^{\infty} S_o(\underline{\omega}) \cos(\underline{\omega} \cdot \underline{\xi})\, d\underline{\omega} = R_o(\underline{\xi}) \tag{27}$$

where it is assumed $S_o(\underline{\omega}) = 0$ for $\underline{\omega} < \underline{\omega}_\ell$ and $\underline{\omega} > \underline{\omega}_u$.

This indicates that, when the ensemble average is considered, the simulated process $f(\underline{x})$ possesses the target autocorrelation $R_o(\underline{\xi})$ and therefore the target spectral density $S_o(\underline{\omega})$.

It can also be shown[2] that the temporal (or spatial) mean $< f(\underline{x}) >$ is zero and the temporal autocorrelation $R^*(\underline{\xi}) = < f(\underline{x})\, f(\underline{x} + \underline{\xi}) >$ becomes

$$R^*(\underline{\xi}) = \sum_{k=1}^{N} A^2(\underline{\omega}_k) \cos(\underline{\omega}_k \cdot \underline{\xi}) \tag{28}$$

As $N \to \infty$, Eq.28 becomes

$$R^*(\underline{\xi}) = \int_{-\infty}^{\infty} S_o(\underline{\omega}) \cos(\underline{\omega} \cdot \underline{\xi})\, d\underline{\omega} = R_o(\underline{\xi}) \tag{29}$$

From Eqs.26 and 28 it is seen that the process $f(\underline{x})$ in Eq.23 is ergodic regardless of the size of N. This makes the method directly applicable to a time domain analysis in which the ensemble average can be evaluated in terms of the temporal average.

Note that the simulated process is Gaussian by virtue of the central limit theorem.

It is important to note that the FFT technique can and

should be used in the digital simulation of multidimensional Gaussian processes writing Eq.24 in a similar form to Eq.8[17].

As an example of digital simulation of a multidimensional process, consider a two-dimensional homogeneous Gaussian process $f_o(t,x)$ with mean zero and spectral density

$$S_o(\omega,k) = \frac{KL^2}{2\pi^2} \cdot \frac{|\omega|}{(1 + c^2\ \omega^2)^{4/3}} \cdot \frac{\alpha|\omega|}{\pi(\alpha^2\omega^2 + k^2)} \qquad (30)$$

where t and x represent the time and the distance respectively and, correspondingly, ω and k the frequency and the wave number. It is known that such process $f_o(t,x)$ is a satisfactory model of a fluctuating part of wind velocity along a straight line direction x. In the wind study, however, it is customary to consider the Fourier transform $S_o(\omega,\xi)$ of the autocorrelation $R_o(\tau,\xi) = E\left[f_o(t,x)\ f_o(t+\tau,\ x+\xi)\right]$ only with respect to τ. In fact, the form of $S_o(\omega,\xi)$ consistent with Eq.30 is

$$S_o(\omega,\xi) = \frac{KL^2}{2\pi^2} \cdot \frac{|\omega|}{(1+c^2\ \omega^2)^{4/3}} \cdot e^{-\alpha|\omega||\xi|} \qquad (31)$$

a familiar form for a fluctuating part of wind velocity at the reference altitude of 33 feet where L = 4000 ft., K = surface drag coefficient, α = constant, $C = L/(2\pi U_{33})$ with U_{33} being the mean wind velocity at the reference altitude. For U_{33} = 40 mph, α = 0.02 ft . sec and K = 0.03, the sample functions $\bar{f}(t,x)$ of $f_o(t,x)$ are computed and shown in Fig.5 at $x = 0$, 50 and 200 ft. One can easily see in this example that the correlation almost disappears as the separation ξ increases to 200 ft.

4. MONTE CARLO SOLUTION OF STRUCTURAL DYNAMICS

The preceding method of digital generation of samle functions of a Gaussian process can be used for the Monte Carlo solution of the following structural problems. It is pointed out parenthetically that by adding a constant value m to the sample functions $\bar{f}(t,\underline{x})$ described in the preceding sections, one can generate sample functions of the simulated process $f(t,\underline{x}) + m$ associated with $f_o(t,x) + m$. Note that the mean value of $f_o(t,\underline{x}) + m$ is no longer zero but is equal to m.

(a) The method can be used in the response analysis of a nonlinear structure under random loading if such loading can be idealized as Gaussian homogeneous or Gaussian evolutionary process with constant mean values. In particular, if the modes $\mu_k(\underline{x})$ of the corresponding linear structures are known. the solution $y_o(t,\underline{x})$ is an approximation expanded into a finite series:

$$y_o(t,\underline{x}) = \sum_{k=1}^{K} q_k(t)\ \mu_k(\underline{x}) \tag{32}$$

When Eq.32 is substituted into the governing (nonlinear partial) differential equation(s) of motion, one can usually get a set of simultaneous nonlinear but ordinary differential equations involving the generalized forces of the following form:

$$F_k(t) = \int_D \mu_k(\underline{x})\ f_o(t,\underline{x})d\underline{x} \tag{33}$$

where $f_o(t,\underline{x})$ is the random excitation process and D indicates an appropriate domain of integration. Sample functions $\bar{F}_k(t)$ can then be digitally generated from Eq.33 with $f_o(t,\underline{x})$ replaced by $\bar{f}(t,\underline{x})$:

$$\bar{F}_k(t) = \int_D \mu_k(\underline{x})\ \bar{f}(t,\underline{x})d\underline{x} \tag{34}$$

It goes without saying that $\bar{f}(t,\underline{x}) + m$ has to be used in place of $\bar{f}(t,\underline{x})$ if the excitation process has a non-zero constant mean value since the simple superposition of solutions does not apply in this case because of nonlinearity.

The modes $\mu_k(\underline{x})$ often take the form of sinusoidal or hyperbolic functions or their combinations. Therefore, the integration in Eq.34 can usually be carried out in closed form since $\bar{f}(t,\underline{x})$ is given as a sum of cosine functions. This is one of the significant advantages of the present method of simulation over other existing methods. In fact, if the domain of integration D represents a two or three dimensional space, the numerical integration of Eq.34 will usually become an isurmountable obstacle. Another advantage is that the present method does not require the nonlinearity to be small or moderate, a condition which has to be imposed for standard linearization or perturbation techniques.

Once the sample functions $\bar{F}_k(t)$ are evaluated from Eq.34 then the sample functions $\bar{q}_k(t)$ of $q_k(t)$ can be numerically evaluated from the (simultaneous) nonlinear but ordinary differential equations mentioned above (replacing of course $F_k(t)$ by $\bar{F}_k(t)$ therein). The experience shows that this phase of numerical work is not a serious problem. Finally, the sample function $\bar{y}(t,\underline{x})$ of the solution $y_o(t,\underline{x})$ can be obtained from Eq.32 with $q_k(t)$ replaced by $\bar{q}_k(t)$. The temporal average of $\bar{y}^2(t,\underline{x})$ over a sufficiently long period of time will produce the mean square response in the Monte Carlo sense if the processes involved are ergodic. Otherwise, the ensemble average has to be considered.

Reference 6 represents a typical example dealing with random responses of nonlinear string and plate. The analysis for the nonlinear string is briefly presented here.

It is pointed out, however, that the digital simulation of the random process for this example was performed using the multidimensional equivalent of Eqs.5 and 6.

The differential equation and the boundary conditions for transverse displacement u of a string under initial tension T_o are, in dimensionless form, given by

$$\frac{\partial^2 \bar{u}}{\partial \bar{t}^2} + 2\beta \frac{\partial \bar{u}}{\partial \bar{t}} =$$

$$= 4\left[1 + \frac{AE}{2T_o}\left(\frac{U}{L}\right)^2 \int_0^1 \left(\frac{\partial \bar{u}}{\partial \bar{t}}\right)^2 d\bar{x}\right] \frac{\partial^2 \bar{u}}{\partial \bar{x}^2} + 4\left(\frac{L}{U}\right) \bar{f}(\bar{x},\bar{t}) \tag{35}$$

$$\bar{u}(0,t) = \bar{u}(1,t) = 0 \tag{36}$$

where

$\bar{x} = X/L$ $\quad\quad \bar{u} = u/U$

$\bar{t} = t/2L\sqrt{\frac{\rho}{T_o}}$ $\quad\quad \bar{f} = \frac{f(\bar{x},\bar{t})L}{T_o}$

L = string length

T_o = initial tension

ρ = mass/unit length

$f(\bar{x},\bar{t})$ = force/unit length

β = $\eta L/\sqrt{\rho}\ T_o$

η = coefficient of linear viscous damping

A = cross-sectional area of string

E = elastic modulus of string

U = reference displacement

Assuming an approximate solution consisting of first M

terms of the modes of the (corresponding) linear string,

$$\bar{u}(\bar{x},\bar{t}) = \sum_{n=1}^{M} b_n(\bar{t}) \sin n\pi\bar{x} \tag{37}$$

Substituting Eq.37 into Eq.35 multiplying by sin $m\pi\bar{x}$, and integrating from 0 to 1, one obtains

$$\frac{d^2 b_n}{d\bar{t}^2} + 2\beta \frac{db_n}{d\bar{t}} = - \left[1 + \alpha \sum_{k=1}^{M} k^2 b_k^2\right] \cdot$$

$$\cdot \; (2n\pi)^2 \, b_n + 8\gamma \int_0^1 \bar{f}(\bar{x},\bar{t}) \sin n\pi\bar{x} \, d\bar{x} \tag{38}$$

$$n = 1,2,\ldots,M$$

where

$$\gamma = \frac{L}{U}$$

$$\alpha = \frac{AE\pi^2}{4T_o} \frac{1}{\gamma^2}$$

The above equations can be solved numerically without difficulty if the associated generalized forces

$$F_n(t) = \int_0^1 \bar{f}(\bar{x},\bar{t}) \sin n\pi\bar{x} \, d\bar{x} \tag{39}$$

are given numerically.

This can be accomplished easily since $\bar{f}(\bar{t},\bar{x})$ can be readily simulated in a manner described above. Note that the integral in $F_n(t)$ can be carried out analytically because $\bar{f}(\bar{t},\bar{x})$ is a sum of cosine functions.

The result of the analysis will be shown in a diagram where the root mean square (r.m.s.) of the response is plotted against that of the excitation. Since, however, the statistical fluctuation in the r.m.s. of the simulated excitation process will undoubtedly reflect on the r.m.s. of the simulated response process, both are estimated by means of temporal average although the theoretical value of the r.m.s. for the excitation process is known. For this purpose, Eq.37 is used for the response upon computing $b_n(t)$ and equivalent of Eq.5 for the excitation.

As an example, a nondimensionalized pressure field, $\bar{f}(\bar{x},\bar{t})$ with a nondimensionalized generalized spectral density

$$\bar{\Phi}_o(\bar{\omega}_1,\bar{\omega}_2) = \frac{ac|\bar{\omega}_1|}{\pi^2(a^2+\bar{\omega}_1^2)(c^2\bar{\omega}_1^2+\bar{\omega}_2^2)} \tag{40}$$

is applied to the string. Eq.40 is obtained from the cross-spectral density $\bar{S}_o(\bar{\omega}_1,\bar{\xi}_2)$ of the following form;

$$\bar{S}_o(\bar{\omega}_1,\bar{\xi}_2) = \bar{S}_o(\bar{\omega}_1,0)e^{-c|\bar{\omega}_1||\bar{\xi}_2|} \tag{41}$$

with

$$\bar{S}_o(\bar{\omega}_1,0) = \frac{a\sigma_f^2}{\pi(a^2 + \bar{\omega}_1^2)} \tag{42}$$

A result is obtained for a particular set of values of parameters involved ($\beta = 2\pi/10$, $a = 4\pi$, $c = 0.7$ and $\alpha = 0.5$) and shown in Fig.6 where r.m.s. value of the nondimensional displacement $\bar{u} = u/(0.1L)$ at the midspan is plotted against that of the nondimensional excitation $\bar{f} = L\, f_o(x,t)/T_o$.

The accuracy of the present approach is first checked against the exact solution obtained for the corresponding linear string. Open circles indicate the r.m.s. values based on simulated excitation processes consisting of 100 cosine terms ($N = 100$) and on the temporal average computed over $T = 5$ cycles of fundamental period $2L\sqrt{\rho(T_o)}$ of the linear string. Open triangles, however, indicate the values based on $N = 500$ and $T = 50$.

It is evident from the computation for the linear string that a use of $N = 500$ and $T = 50$ produces a reasonable approximation. In fact, the r.m.s. value of the nonlinear string (dashed curve) is obtained by interpolating those points (solid triangles) computed with $N = 500$ and $T = 50$. In the present example, the displacement is dominated by the first mode, both in linear and nonlinear cases, occupying approximately 98 % of the total displacement at the midspan. Finally, it is pointed out that fewer number of cosine terms would be needed if Eq.23 was used instead of equivalent of Eq.5 for the simulation of the excitation process $\bar{f}(\bar{t},\bar{x})$.

The application of this type of Monte Carlo approach

has also been made to other nonlinear structural response analysis[1,2,5,7,27].

(b) The method can be applied to the failure analysis of a structure with spatially random variation of strength and other material properties. In this case, sample structures are generated by digitally generating such spatial variations of strength and other material properties. When correlated spatial variations are observed on more than one material property (e.g. Young's modulus and density), usually a multidimensional multivariate process has to be generated.

Applying to each of these sample structures a sample stress history of a random stress process, the fatigue life of a sample structure can be computed under the assumption of a certain fatigue failure mechanism. The statistical variation of the fatigue life thus computed establishes its empirical distribution under the random stress process in the Monte Carlo sense. This approach was successfully taken in Ref.28. A similar problem in which the empirical distribution of the static failure load is to be found for a concrete structure with spatial strength variation is treated in detail in Refs.13 and 19.

(c) The method can be employed effectively when the structural system to be analyzed is complex even though it involves neither nonlinearity nor random variation of material properties. The mean square response (displacement, shear force and bending moment) of a large floating plate to wind-induced random ocean-waves are computed in Ref.29 taking the temporal averages of sample response functions as in Ref.6. The analysis is essentially numerical since sample functions of the wind-induced ocean-surface elevation are digitally generated and the corresponding response functions are numerically obtained. This was done because the ocean-structure system considered was too complex to solve analytically. Another example of this kind is the study of the dynamic interaction between moving vehicles and a bridge with random pavement surface roughness[8]. In this problem the random pavement surface roughness is digitally simulated for numerical response analysis.

In some problems of mechanics, a structure is considered complex when its material properties are spatially random. The wave propagation through a random medium is one of these problems. In Ref.14, the stress wave propagation through a finite cylinder with random material properties is treated under the condition that the one end of the cylinder is acted upon by an impact load and the other end is free. A set of one hundred samples of correlated random material properties (Young's modulus and density) are generated thus producing one hundred sample cylinders. The finite element method is applied for the stress analysis to compute maximum stress intensity in each of these cylinders due to the impact. An empirical distribution of the maximum stress intensity is then established in the Monte Carlo sense.

(d) Finally, the method is often useful when the problem is to determine eigenvalues (frequencies and buckling loads) of the structure with random material properties. As in the case of the wave propagation problems, sample structures are generated and the statistical distribution of eigenvalues of these structures are treated as the empirical distribution of the eigenvalue of interest. An example of this problem is given in Ref.9.

REFERENCES

[1] Shinozuka, M., Simulation of multivariate and multidimensional random processes, J. Ac. Soc. Am., 49, 1 (part 2), 357, January 1971.

[2] Shinozuka, M. and Jan C.-M., Digital simulation of random processes and its applications, J. Sound and Vibration, 25, 1,111, 1972.

[3] Yang, J.-N., Simulation of random envelope processes, J. Sound and Vibration, 21, 73, 1972.

[4] Shinozuka, M. and Sato, Y., Simulation of nonstationary random processes, Proc.ASCE, EMD Journal, 93, EM 1, 11, 1967.

[5] Shinozuka, M. and Wen Y.-K., Nonlinear dynamic analysis of offshore structures; a Monte Carlo approach, Proc. of the Int. Symp. on Stoch. Hydraulics, Pittsburgh, 507, June 1971.

[6] Shinozuka, M. and Wen,Y.-K. Monte Carlo solution of nonlinear vibrations, AIAA J. 10, 37, January 1972.

[7] Wen, Y.-K. and Shinozuka, M., Monte Carlo solution of structural response to wind load, Proc. of the 3rd Int. Conf. on Wind Effects on Buildings and Structures, Tokyo, Japan, Part III, 3-1, Sept.1971.

[8] Shinozuka, M. and Kobori T., Fatigue analysis of highway bridges, Proc. of Japan Socity of Civil Engineers, 208, 137, Dec. 1972.

[9] Shinozuka, M. and Astill, J.C., Random eigenvalue problems in structural mechanics, AIAA J., 10, 4, 456, April 1972.

[10] Vaicaitis, R. and Shinozuka, M., Simulation of flow induced vibration, ASCE National Structural Engineering Meeting, Cincinnati, Meeting Preprint 2218, April 1974.

[11] Shinozuka, M., Monte Carlo solution of structural dynamics, Int. J. of Computers and Structures, 2, 855, 1972.

[12] Vaicaitis, R. and Shinozuka, M. and Takeno, M., Parametric study of wind loading on structures, Proc.ASCE, J.STD, 99, ST 3, 453, March 1973.

[13] Shinozuka, M., Probabilistic formulation for analytical modeling of concrete structures, Proc. ASCE, M. EMD, 98, EM 6, 1433, December 1972.

[14] Astill, C.J. and Nosseir, S.B. and Shinozuka, M., Impact loading on structures with random properties, J. of Struct. Mech., 1, 1, 63, 1972.

[15] Vaicaitis, R. and Shinozuka, M. and Takeno, M., Response analysis of tall buildings, Proc. ASCE, J.STD, 101, ST 3, 585, March 1975.

[16] Shinozuka, M., Digital simulation of ground acceleration, 5-th WCEE, Rome, Italy, 2829, June 1973.

[17] Shinozuka, M., Digital simulation of random processes in engineering mechanics with the aid of FFT technique in Stochastic Problems in Mechanics, Ariaratnam, S.T. and Leipholz, H.H.E.,University of Waterloo Press, 277, 1974.

[18] Wilkins, D.J. and Wolff, R.V. and Shinozuka, M. and Cox, E.F., Realism in fatigue testing, the effect of flight-by-flight thermal and random load histories on composite bonded joints, ASTM, STP 569, 307, 1975.

[19] Shinozuka, M. and Lenoe, E., A probabilistic model for spatial distribution of material properties, J. of Eng. Fracture Mech., 8, 217, 1976.

[20] Shinozuka, M. and Levy R., Generation of wind velocity field for design of antenna reflectors, accepted for publication in Eng. Mech. Div. Proc.ASCE.

[21] Shinozuka, M. and Vaicaitis, R. and Asada, H., Digital simulation of random forces for large scale experiments, accepted for publication in AIAA J.

[22] Rice, S.O., Mathematical analysis of random noise, in Selected Papers on Noise and Stochastic Processes edited by N.Wax, Dover Publications, Inc., New York, 180, 1954.

[23] Borgman, L.E., Ocean wave simulation for engineering design, Proc.ASCE, J. of Waterway and Harbors Div., 95, WW 4, 556, Nov. 1969.

[24] Cramer, H. and Leadbetter, M.F., Stationary and Related Stochastic Processes, John Wiley, New York, 249, 1967.

[25] Yang, J.-N., On the normality and accuracy of simulated random processes, J. Sound and Vibration, 26, 3, 417, 1973.

[26] Bochner, S., Lectures on Fourier Integrals, English Translation by M. Tenebaum and Pollard H., Annals of Mathematic Studies, 42, Princeton University Press, 325, 1959.

[27] Vaicaitis, R. and Jan C.-M. and Shinozuka, M., Non-linear panel response and noise transmission from a turbulent boundary layer by a Monte Carlo approach, AIAA J., 10, 7, 895, July 1972.

[28] Itagaki, H. and Shinozuka, M., Application of Monte Carlo Technique to Fatigue Failure Analysis under Random Loading, ASTM, STP 511,168.

[29] Wen, Y.-K. and Shinozuka, M., Analysis of Floating Plate under Ocean Waves, Journal of Waterways, Harbors and Coastal Engineering Division, Proc. ASCE, 98, WW2, 177, May 1972.

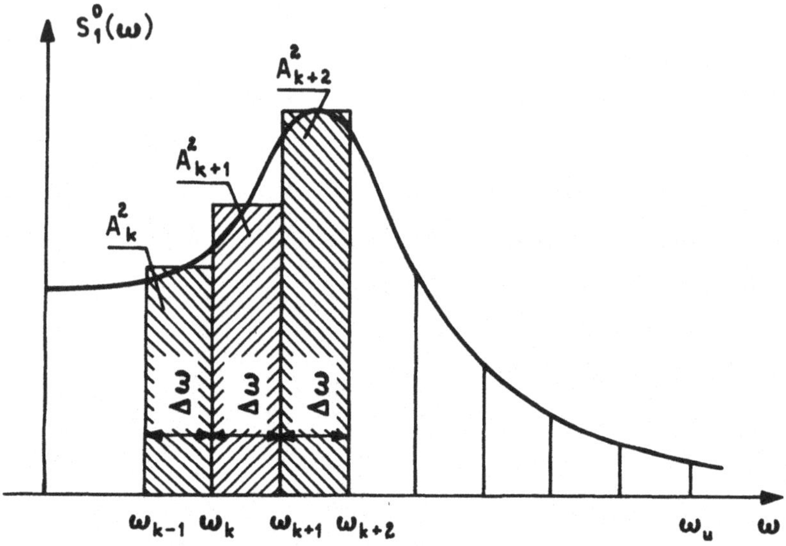

$$\bar{f}(x) = \sqrt{2}\sum_{k=1}^{N} A_k \cos(\omega_k x - \phi_k)$$

$$\text{where} \quad A_k \doteq \sqrt{S_1^0(\omega_k)\,\Delta\omega}$$

$$\omega_k = k\Delta\omega$$

$$\omega_u = N\Delta\omega$$

Fig. 1 One-Sided Spectral Density

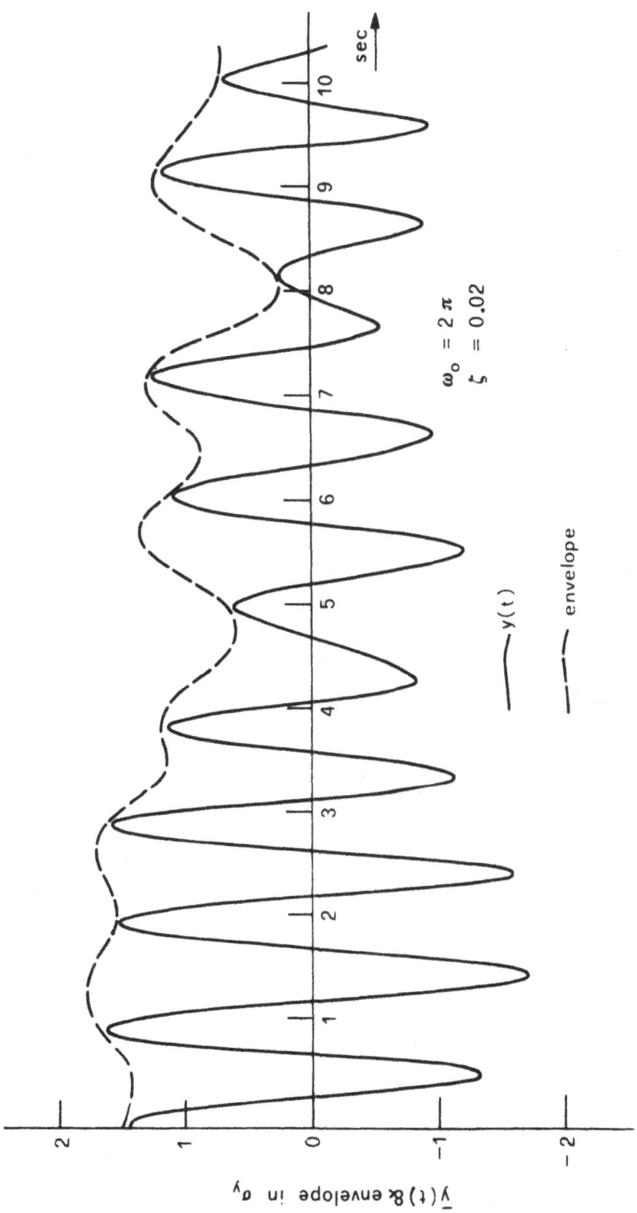

Fig. 2 Sample Functions of a Narrow-Band Random Process and its Envelope Process

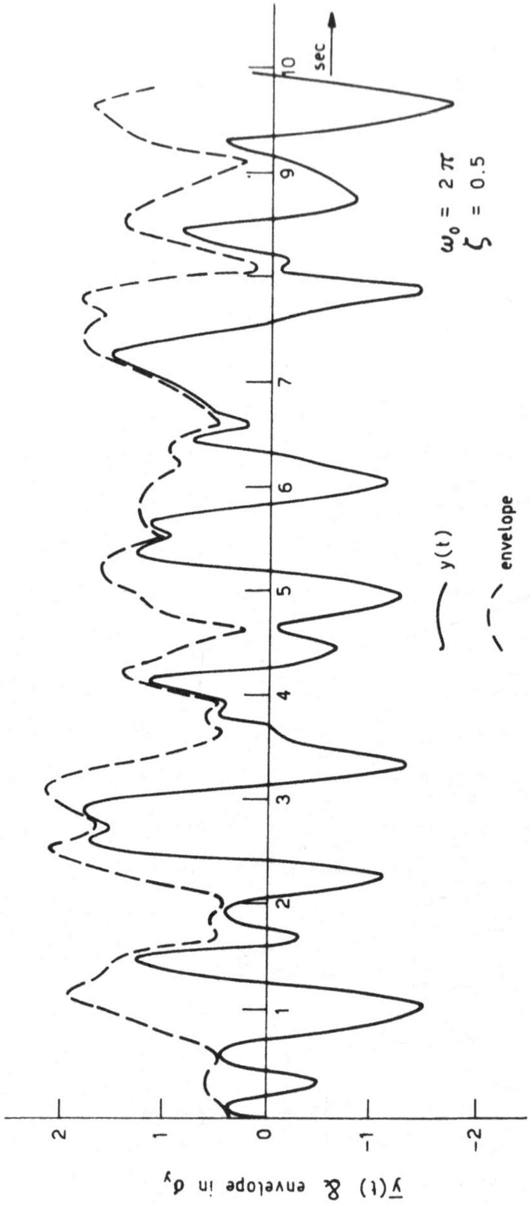

Fig. 3 Sample Functions of a Wide-Band Random Process and its Envelope Process

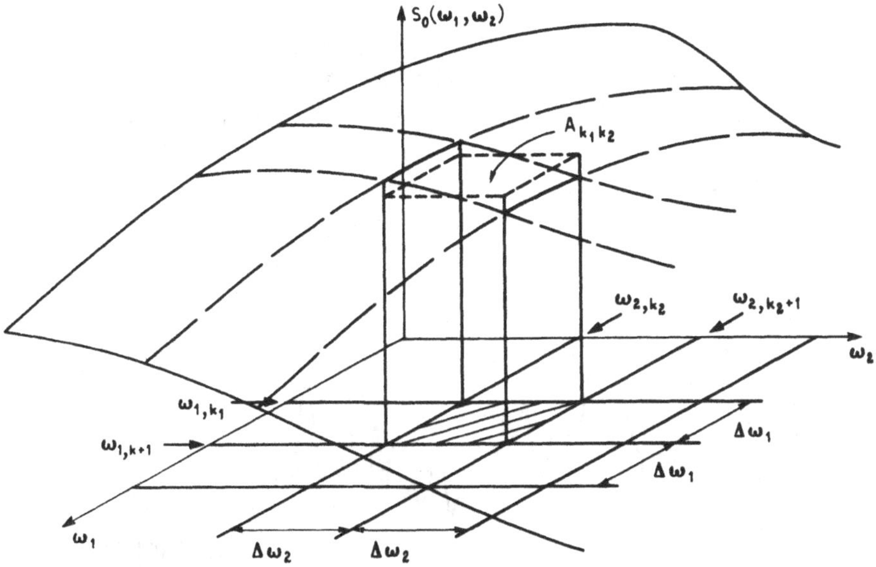

$$\bar{f}(x_1,x_2) = \sqrt{2}\sum_{k_1=1}^{N_1}\sum_{k_2=1}^{N_2} A_{k_1k_2}\cos(\omega_{1k_1}x_1 + \omega_{2k_2}x_2 - \phi_{k_1k_2})$$

where $A_{k_1k_2} \doteq \sqrt{S_O(\omega_{1k_1},\omega_{2k_2})\,\Delta\omega_1\,\Delta\omega_2}$

$$\omega_{1k_1} = k_1\Delta\omega_1,\ \omega_{2k_2} = k_2\Delta\omega_2$$

Fig. 4 Two-dimensional Spectral Density

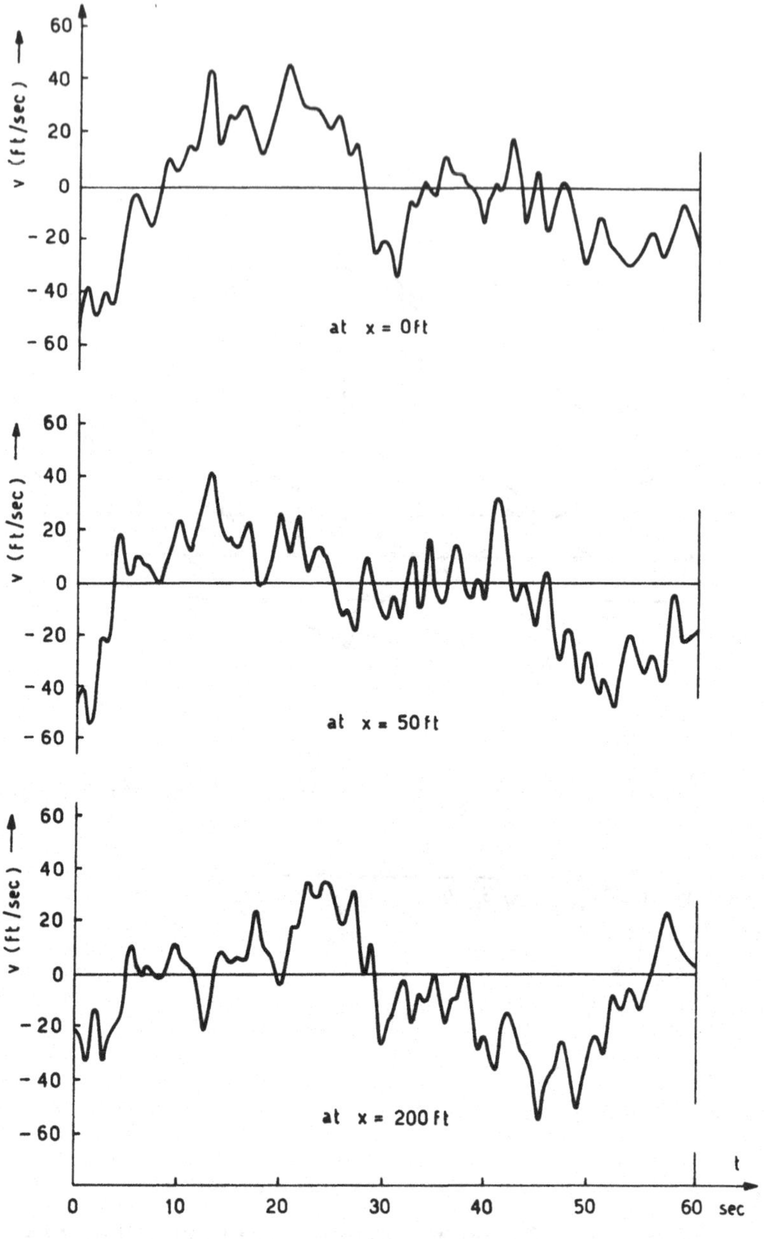

Fig. 5 Simulation of Wind Velocity at Different Points

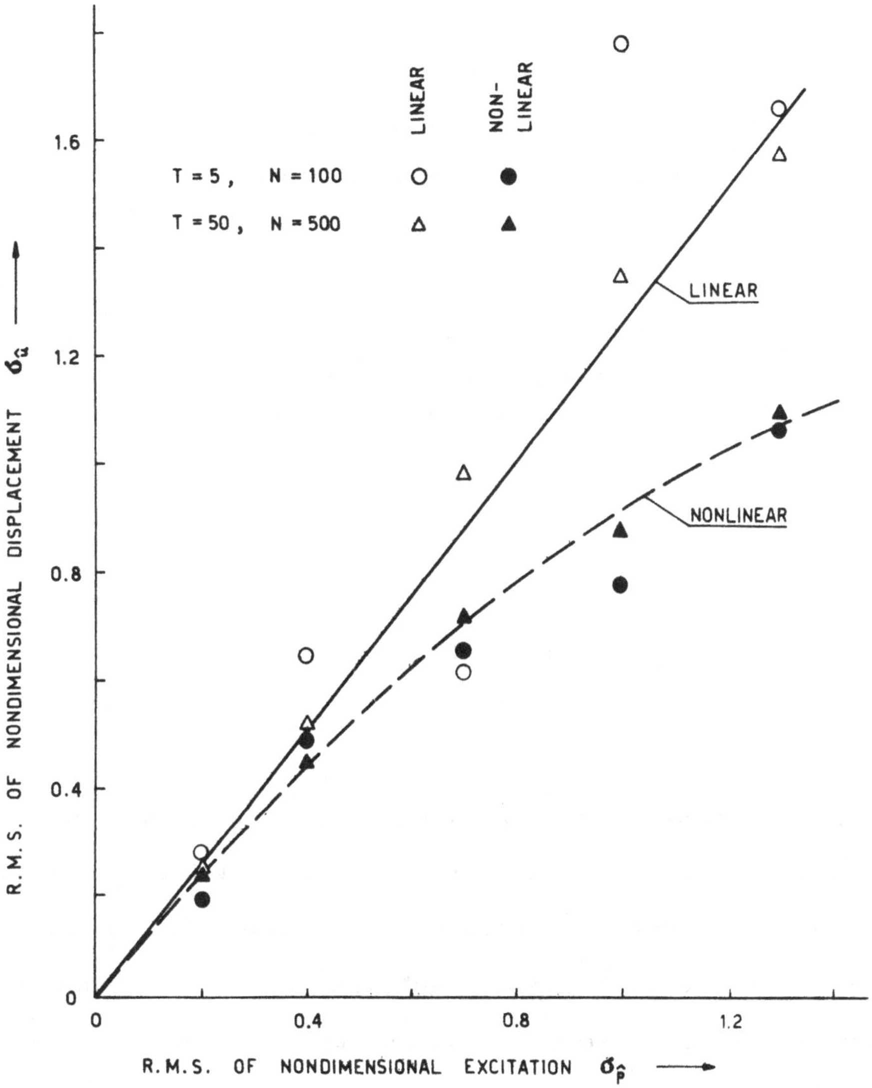

Fig. 6 R.M.S. Response of Non-Dimensional Displacement (String)

STRUCTURAL RESPONSE UNDER TURBULENT FLOW EXCITATIONS

Y.K. LIN
University of Illinois
Urbana, Illinois, U.S.A.

ABSTRACT

In this report three problems of random vibration are discussed. In each case the external excitation is a turbulent flow. The first two problems, an airplane flying into atmospheric turbulence and a panel-like structure exposed to boundary-layer pressure fluctuation, are treated as linear problems. It is shown that if Taylor's hypothesis of a frozen turbulence field is valid then the calculation can be greatly simplified using a spectral analysis in the wave-number domain. However, if decay in the turbulence is appreciable a superposition scheme can still be used to retain as much computational advantage of the wave-number domain analysis as possible.

The third problem, the response of a building to gusty wind, is formulated as a nonlinear problem in which randcm inputs occur both as parametric and non-parametric excitations. The stochastic averaging method of Stratonovich and Khasminskii is used to obtain equivalent Itô equations for the along-wind motion and the across-wind motion. Stability conditions are established for the second moment in the along-wind direction and for the first moment in the across-wind direction. The stationary second moment for the along-wind motion, when it is stable, is also obtained.

ACKNOWLEDGEMENTS

This work was carried out when the author was a Senior Visiting Fellow at the Institute of Sound and Vibration Research of the University of Southampton. The financial support from the Science Research Council of the United Kingdom is gratefully acknowledged. The author wishes to thank Professor B.L. Clarkson of the Institute for unfailing faith and constant encouragement and Dr.P.J. Holmes for many valuable technical discussions. The preliminary version of this report has been presented in a series of ten lectures at the International Centre for Mechanical Sciences at Udine, Italy, from 12 to 16 July 1976.

I. INTRODUCTION

In these lecture notes three problems of practical interest will be presented. The first problem is the response of an airplane to atmospheric turbulence. The second is concerned with a panel-like structure exposed to boundary-layer turbulence. The last is an exploratory analysis of building response to gusty wind. The configuration of the structure in each case is kept simple, so that the main features of the treatment will not be obscured by computational complexities. The first two problems will be treated as linear, but in the last problem non-linear and stability analyses will be presented. Much of the materials are generated from the writer's own work, some of which are yet to be published at the time of these lectures.

To set the stage for discussions, let us refer to Fig.1.1 which shows the general nature of a fluid-structure interaction problem. The fluid flow generates a force field F(x,t) on the structure and causes the deflection W(x,t) of the structure, links (1) and (2). The structural motion, however, will change the force field as well as the velocity field U(x,t), links (3) and (4). Therefore, the problem is clearly non-linear. However, no analysis is known to date which accounts for all the four links, especially link (4). To make a problem simpler and consequently solvable, one often uses the linearization scheme shown in Fig.1.2. In this case, F_o represents the forcing field if the structure is kept either motionless, or in a reference steady state, and F_1 is the additional forcing field generated by the structural motion, or perturbed motion from the reference stae. The calculation of F_1 is based on the assumption that it is not influenced by the presence of F_o.

Our analysis for problems one and two will follow the scheme of Fig.1.2. Problem three will be treated in the manner of Fig.1.1, but minus link (4).

II. AIRPLANE RESPONSE TO ATMOSPHERIC TURBULENCE

The simplest model for the study of this case is shown in Fig.2.1. The following is a list of assumptions:

(1) The airplane is rigid and the only significant perturbation from its steady forward flight is the up-and-down plunging motion.

(2) The forward speed u is a constant and is directed toward the $-x_1$ direction.

(3) The turbulence field is considered to be frozen and statistically homogeneous in space (Taylor's hypothesis).

(4) Among all three components of the turbulent (perturbation) velocity, only the vertical component V_2 affects the motion of the vehicle. With the above assumptions, it is sufficient to represent the flow velocity field relative to the vehicle as

$$U_i = u\,\delta_{i1} + V\,\delta_{i2} \tag{2.1}$$

Clearly the steady speed u is required to maintain the steady lift. We shall be interested in the vertical acceleration Z(t) of the airplane, the response quantity of interest.

Since the gust field is assumed to be frozen in space, the airplane senses a time-varying excitation which is a function of $(x_1 - ut)$. Such a turbulence field can be represented by a Stieltjes-Fourier integral

$$V(x_1 - ut) = \int_{-\infty}^{\infty} e^{ik(x_1 - ut)} d\,\tilde{V}(k) \tag{2.2}$$

This equation implies that V is composed of infinitely many frozen-pattern sinusoids. For simplicity, we shall drop the subscript 1 associated with x in what follows.

Within the framework of a linear analysis, we can also construct the total vehicle response to an arbitrary frozen random velocity field from a basic solution corresponding to a unit frozen sinusoid velocity field. In a certain sense, this basic solution is analogous to the well-known frequency response function. Thus we consider first the problem shown in Fig.2.2. The major part of this problem actually has been considered by Sears[1] for the case of imcompressible fluid. Specifically, Sears considered the change of lift on a two-dimensional airfoil flying into a sinusoidal gust. The result is

$$\eta_1(k) = \sqrt{2}\,\pi\,\rho\,bu\,\phi(k) \tag{2.3}$$

where ρ is the air density, b is one-half of the chord-width of the lifting surface and $\phi(k)$ is the now well-known Sears' function. The exact Sears' function involves Bessel's functions, but Liepmann[2] has suggested an approximation:

$$|\phi(k)|^2 = \frac{1}{1 + 2\pi|kb|} \tag{2.4}$$

For certain purposes this approximation leads to unbounded

results. Other approximations include one by Howell and Lin[3] as follows:

$$\phi(k) = \left(\frac{0.065}{0.13 + ikb} + \frac{0.5}{1 + ikb}\right) e^{1.10\, ikb} \tag{2.5}$$

Additional adjustments are required for compressible flow, but the general form remains similar to (2.5).

Another transformation is required to transfer the change of lift to the vehicle acceleration response for which we require[4]

$$\eta_2(k) = \sqrt{2}\ kb \{\pi o A b \left[kb(1 + 2\lambda) - 2i\right]\}^{-1} \tag{2.6}$$

where, in addition to those previously defined, A is the total lifting surface area, and λ is a mass parameter, given by

$$\lambda = \frac{m}{\pi \rho A b} \tag{2.7}$$

and m the total mass of the aircraft.

It is to be noted at this point that both transfer functions η_1 and η_2 are expressed in terms of wavenumber k, which, referring to equation (2.2), is related to the frequency ω, as

$$ku = \omega \tag{2.8}$$

in the case of a frozen random field. The product kb is, in

fact, the well-known Strouhal's number

$$s = kb = \frac{\omega b}{u}$$

In the terminology of aero-elasticity, s is often called the reduced frequency.

The frequency response function from gust velocity to airplane acceleration is then

$$H(\omega) = \eta_2(\frac{\omega}{u})\ \eta_1(\frac{\omega}{u}) \qquad (2.9)$$

and the acceleration response at the stationary state is

$$Z(t) = \int_{-\infty}^{\infty} H(\omega) e^{-i\omega t} d\tilde{V}(\frac{\omega}{u}) \qquad (2.10)$$

However, it is not necessary to convert k to ω in actual calculations. Clearly Z(t) can also be computed from

$$Z(t) = \int_{-\infty}^{\infty} \bar{H}(k) e^{-ikut} d\tilde{V}(k) \qquad (2.11)$$

where $\bar{H}(k) = \eta_2(k)\eta_1(k)$.

It is known from the theory of random processes that $\tilde{V}(k)$ is a random process of uncorrelated increments; that is,

$$\left[\tilde{V}(k_2) - \tilde{V}(k_1)\right], \quad \left[\tilde{V}(k_4) - \tilde{V}(k_3)\right]$$

are uncorrelated if $k_1 < k_2 \leq k_3 < k_4$. The first and the second order statistical properties of the $\tilde{V}(k)$ process are similar to the Brownian process which is one of independent increments. Therefore, $\tilde{V}(k)$ does not have a mean-square derivative. In fact, $d\tilde{V}(k)$ has a magnitude of the order of $(dk)^{1/2}$

and the reason for representing (2.2) as a Stieltjes-Fourier integral instead of the usual Riemann-Fourier integral is now clear. However, $d\tilde{V}(k)$ is related to the spectral density of V. Specifically,

$$E\{d\tilde{V}(k_1)d\tilde{V}^*(k_2)\} = \begin{cases} 0, & k_1 \neq k_2 \\ S_v(k)dk, & k_1 = k_2 = k \end{cases} \tag{2.12}$$

where $S_v(k)$ is called the spectral density of V in the wave-number domain.

We now can calculate the correlation function of the response $R_Z(\tau) = E[Z(t)Z(t+\tau)]$. We replace k by k_1 in equation (2.11), and write another expression for $Z(t+\tau)$ but in the complex conjugate form and with k replaced by k_2. Take the ensemble average of the product. We obtain

$$R_Z(\tau) = \int_{-\infty}^{\infty} |\bar{H}(k)|^2 \, e^{iku\tau} S_v(k)dk \tag{2.13}$$

This has a frequency domain analog of

$$R_Z(\tau) = \int_{-\infty}^{\infty} |H(\omega)|^2 \, e^{i\omega\tau} \, \Phi_v(\omega)d\omega \tag{2.14}$$

where $\Phi_v(\omega)$ is, of course, the spectral density of V in the frequency domain. Comparing (2.13) and (2.14), it is seen that

$$\Phi_v(\omega) = \frac{1}{|u|} S_v\left(\frac{\omega}{u}\right) \tag{2.15}$$

In terms of spectral density of the response, we also

have the two form:

$$S_Z(k) = |\bar{H}(k)|^2 S_v(k)$$ (2.16)

and

$$\phi_Z(\omega) = |H(\omega)|^2 \phi_v(\omega)$$

$$= \frac{1}{|u|} |H(\omega)|^2 S_v(\frac{\omega}{u})$$ (2.17)

It may be helpful conceptually to regard $\phi_v(\omega)$ as the "fixed frame" frequency spectrum, "fixed" in the sense of fixed coordinates on the airplane, and to regard $S_v(k)$ as the moving frame wave-number spectrum, "moving" in the sense of moving coordinates with respect to the airplane. But, of course, the airplane is actually moving whereas the turbulence field is assumed to be frozen in space.

Several different expressions have been used for the spectrum of a homogeneous gust field in practical applications. The so-called Dryden's spectrum is given by

$$S_v(k) = \frac{\sigma^2 L}{2\pi} \cdot \frac{1 + 3(kL)^2}{\left[1 + (kL)^2\right]^2}$$ (2.18)

where σ^2 is the variance of the turbulence (which has a zero mean), and L is called the scale of turbulence which is a measure of the correlation length in space. Another is the von Karman spectrum which differs from (2.18) in that it tends to zero as $k^{-5/3}$ instead of k^{-2} for large k.

At this point it is of interest to point out several extensions of the above analysis. First, if the airplane is modelled as a multi-degree-of-freedom system, but keeping the assumption that only the vertical component of the turbulence field affects the motion of the airplane, then we require the cross-spectrum of the vertical turbulence velocity, for which we quote the work by Lin[5]

$$S_v(x_1,y_1;\ x_2,y_2;\ k)$$

$$= \frac{2\sigma^2}{\pi}\exp(-ik\xi)\left\{k^2\left[\frac{\eta^2}{4a^2}K_2\left(a\frac{\eta}{L}\right)\right]\right.$$

$$\left.+\left(\frac{1}{2a}\right)^2\frac{a\eta}{L}\left[3K_1\left(a\frac{\eta}{L}\right)-\frac{a\eta}{L}K_2\left(a\frac{\eta}{L}\right)\right]\right\} \qquad (2.19)$$

where $\xi = x_1 - x_2$, $\eta = y_1 - y_2$, $a = \left[1 + (Lk)^2\right]^{1/2}$, and K_m are modified Bessel functions of the second kind. As $\xi = \eta = 0$, (2.19) reduces to (2.18) since

$$zK_1(z) \to 1, \qquad z^2K_2(z) \to 2$$

Another extension is to model the turbulence field as being an inhomogeneous frozen pattern in space. Among several contributions we quote the works by Howell and Lin[3] for the single-degree-of-freedom case and Fujimori and Lin[6] for the multi-degree-of-freedom case.

III. RESPONSE OF A PANEL-LIKE STRUCTURE TO BOUNDARY-LAYER TURBULENCE

Some basic differences exist between the panel problem considered in the following and the airplane problem discussed in Section II. First of all, since the fluid velocity relative to the structure must be zero at the interface, all experimental measurements for the purpose of computing structural response have been directly aimed at the pressure fluctuation itself. Secondly, it is not acceptable to assume the pressure field as being a frozen pattern.

As is seen from equation (2.19), the cross-spectrum of a truly frozen pattern field must have the form that the dependence in the spatial separation ξ in the direction of convection appears only in the imaginary exponential; that is, it must appear as $\exp(-ik\xi)$ in the wave-number domain or as $\exp(-i\omega\xi/u)$ in the frequency domain where u is now the convection velocity. This dependence on the spatial separation can be proved generally, not restricted to the special case of equation (2.19). However, in experimental measurements it has been found that the frequency domain cross-spectrum can be fitted nicely in the general form of

$$\overline{\phi}_p(\xi,\omega) = \overline{\phi}_p(0,\omega)\psi(\xi)\exp(-i\omega\xi/u_c) \qquad (3.1)$$

known as Corcos' form who appears to be the first to have
used such an expression[7]. In equation (3.1) $\overline{\phi}_p(0,\omega)$ is, of
course, the pressure spectrum at any point (the pressure is
assumed implicitly as being homogeneous), and $\psi(\xi)$ is positive definite, and it has an absolute maximum equal to unity
at $\xi = 0$. For a frozen-pattern pressure field, $\psi(\ )$ would
reduce to the constant one.

Even though the structure to be considered has an infinite number of degrees of freedom, the assumption of a
frozen-pattern forcing field, if acceptable, would result
in tremendous computational advantage. In fact, we could
then find a relation between the input and the output similar to equation (2.16) or equation (2.17). To take advantage
of this benefit as much as possible, Lin and Maekawa[8] recently have proposed the following superposition scheme

$$p(x,t) = \int_{-\infty}^{\infty} \hat{p}\,(x - ut)\,dG(u) \qquad (3.2)$$

The implication of (3.2) is clear. The total pressure field
$p(x,t)$ is assumed to be a superposition of infinitely many
small frozen-pattern components, convected in two opposite
directions and at different convection velocities.

Now each frozen component can, again, be decomposed in
the manner of equation (2.2). Therefore,

$$p(x,t) = \int_{-\infty}^{\infty} \int e^{i(u\beta t - \beta x)}\,dF(\beta,u)\,dG(u) \qquad (3.3)$$

and the cross-correlation of p(x,t) is given by

$$E\left[p(x_1,t_1)p(x_2,t_2)\right] = \iiiint_{\infty}^{\infty} e^{i(u_1\beta_1t_1-u_2\beta_2t_2)}$$

$$e^{-i(\beta_1x_1-\beta_2x_2)} E\left[dF(\beta_1,u_1)dF^*(\beta_2,u_2)dG(u_1)dG^*(u_2)\right] \quad (3.4)$$

In order that the right hand side of (3.4) can become only a function of $x_1-x_2 = \xi$ and $t_1-t_2 = \tau$, so that p(x,t) is, indeed, stationary in time and homogeneous in x, it is necessary that

$$E\{dF(\beta_1,u_1)dF^*(\beta_2,u_2)dG(u_1)dG^*(u_2)\}$$

$$=S_p(\beta_1,u_1)\delta(\beta_1-\beta_2)\delta(u_1-u_2)d\beta_1d\beta_2du_1du_2 \quad (3.5)$$

Of course, the delta functions in (3.5) is just a formal way to write a relationship of the type of (2.12). Substituting (3.5) into (3.4), we obtain

$$R_p(\xi,\tau) = \iint_{-\infty}^{\infty} e^{i(\beta u\tau-\beta\xi)} S_p(\beta,u)d\beta\ du \quad (3.6)$$

A Fourier transformation results in

$$\phi_p(\xi,\omega) = \frac{1}{2\pi} \int_{-\infty}^{\infty} R_p(\xi,\tau)e^{-i\omega\tau} d\tau$$

$$= \int_{-\infty}^{\infty} \frac{1}{|u|} e^{-i\omega\xi/u} S_p(\frac{\omega}{u},u)du \quad (3.7)$$

Equation (3.7) is a theoretical pressure cross-spectrum derived under the assumption that the superposition scheme (3.2) is valid.

We now wish to compare (3.7), the theoretical cross-spectrum with the experimental one, (3.1). First we apply a Fourier transform to (3.1), resulting in

$$\frac{1}{2\pi}\int_{-\infty}^{\infty}\bar{\Phi}_p(\xi,\omega)e^{i\xi\alpha}d\xi = \bar{\Phi}_p(0,\omega)\Psi(\alpha - \frac{\omega}{u_c}) \tag{3.8}$$

where

$$\Psi(v) = \frac{1}{2\pi}\int_{-\infty}^{\infty}\psi(\xi)e^{iv\xi}d\xi \tag{3.9}$$

Therefore, the experimental cross-spectrum has a representation of

$$\bar{\Phi}_p(\xi,\omega) = \bar{\Phi}_p(0,\omega)\int_{-\infty}^{\infty}\Psi(\alpha - \frac{\omega}{u_c})e^{-i\alpha\xi}\,d\alpha$$

Let $\alpha = \frac{\omega}{u}$, we arrive at

$$\bar{\Phi}_p(\xi,\omega) = \bar{\Phi}_p(0,\omega)\int_{-\infty}^{\infty}\left|\frac{\omega}{u^2}\right|\Psi(\frac{\omega}{u} - \frac{\omega}{u_c})e^{-i\xi\omega/u}du \tag{3.10}$$

Equating (3.10) and (3.7):

$$\begin{aligned} S_p(\frac{\omega}{u},u) &= \left|\frac{\omega}{u}\right|\bar{\Phi}_p(0,\omega)\Psi(\frac{\omega}{u} - \frac{\omega}{u_c}) \\ &= \left|\frac{\omega}{u}\right|\bar{\Phi}_p(0,\omega)\left\{\frac{1}{2\pi}\int_{-\infty}^{\infty}\psi(\xi)e^{i\xi(\frac{\omega}{u} - \frac{\omega}{u_c})}d\xi\right\} \end{aligned} \tag{3.11}$$

Equation (3.11) provides a procedure whereby a measured cross-spectrum, characterized by $\bar{\Phi}_p(0,\omega)$, the spatial decay function $\psi(\xi)$, and a dominant convection speed, can be incorporated into the theoretical spectrum.

We shall show why the theoretical form (3.7) for the pressure cross-spectrum has a great advantage in the structural response computations. By a generalization of equation (2.17) to the case of a continuous structure (restricted to one-dimensional spatial coordinate for simplicity), we obtain

$$\Phi_W(x_1,x_2;\omega) = \int_{-\infty}^{\infty} \frac{1}{|u|} H(x_1,\omega)H^*(x_2,\omega)S_p(\frac{\omega}{u},u)du \qquad (3.12)$$

The integration variable u can now be converted to the wave-number k, while keeping ω fixed. We note that

$$\frac{\omega}{u} = k, \qquad u = \frac{\omega}{k}, \qquad du = \left|\frac{\omega}{k^2}\right| dk$$

Consequently,

$$\Phi_W(x_1,x_2;\omega) = \int_{-\infty}^{\infty} \frac{1}{|k|} \bar{H}(x_1,k)\bar{H}^*(x_2,k)\ S_p(k,\ \frac{\omega}{k})dk \qquad (3.13)$$

Substituting (3.11) into (3.13) we obtain an extremely simple result

$$\Phi_W(x_1,x_2;\omega) = \bar{\Phi}_p(0,\omega) \int_{-\infty}^{\infty} \bar{H}(x_1,k)\bar{H}^*(x_2,k)\Psi(k - \frac{\omega}{u_c})dk \qquad (3.14)$$

The simplicity of equation (3.14) may not be fully appreciated by casual readers since it still remains in an integral form. However, if one recalls that for spectral analysis of a one-dimensional structure using the conventional distributed load representation, one would require a double integration over spatial coordinate. Replacing it by a single integration should result in several orders of magnitude reduction in computer time. The fact that (3.14) is an infinite integral is not a problem. The wave-number response function, $\bar{H}(x,k)$, diminishes quickly at large absolute values of k. Furthermore, the range of ω that needs to be considered is restricted within the dominant spectral range of the input spectrum $\phi_p(0,\omega)$, as is to be expected and clearly demonstrated in equation (3.14).

For the application to a specific dynamic system, one must know the wave-number response function $\bar{H}(x,k)$ of that system. This wave-number response function is one associated with a unit sinusoidal pressure distribution frozen in space, the computation of which is considerably simpler. Again, we shall choose the simplest model possible which, nevertheless, retains important features of the panel vibration problem for airplanes, ships, or other high speed transportation systems using panel construction on the bodies. The one we choose for the ensuing analysis is an infinite beam shown in Fig.3.1. On the upper side of the beam ($z > 0$) the beam is exposed to a fluid with density ρ_1 and sound speed a_1. On the lower side $z < 0$, it forms one boundary of a cavity which contains a fluid with density ρ_2 and sound speed a_2. Fluid 1 has an ambient velocity u_∞ relative to the beam whereas fluid 2 is initially quiescent with respect to the beam.

The problem of computing $\bar{H}(x,k)$ can be represented symbolically as finding the steady state solution of the following equation

$$\mathcal{L}\left\{\bar{H}(x,k)e^{i\omega t}\right\} = e^{i(\omega t-kx)} \tag{3.15}$$

where $\mathcal{L}$ is a linear operator since the problem is assumed to be linear.

The equation of motion for the beam is

$$EI\frac{\partial^4 w}{\partial x^4} + m\frac{\partial^2 w}{\partial t} = p + (p_1 - p_2)_{z=0} \tag{3.16}$$

For the determination of the wave-number response $\bar{H}$, p is replaced by $\exp\left[i(\omega t - kx)\right]$ and w by

$$w = \bar{H}(x,k)e^{i\omega t} = A(k)\exp\left[i(\omega t - kx)\right] \tag{3.17}$$

The second half of equation (3.17) is derived from the fact that at the steady state the exponential factor $\exp\left[i(\omega t-kx)\right]$ can be cancelled from both sides of the equation.

The acoustic pressure induced by the structural motion on the upper side of the beam, p_1, is governed by the convected wave equation

$$(\frac{\partial}{\partial t} + u_\infty \frac{\partial}{\partial x})^2 p_1 - a_1^2(\frac{\partial^2}{\partial x} + \frac{\partial^2}{\partial z^2}) p_1 = 0 \qquad (3.18)$$

and satisfies the boundary condition

$$(\frac{\partial p_1}{\partial z})_{z=0} = \rho_1(\frac{\partial}{\partial t} + u_\infty \frac{\partial}{\partial x})^2 w \qquad (3.19)$$

and the radiation condition that it propagates only in the domain $z > 0$. Note that the ambient velocity of the fluid u_∞ appearing in equations (3.18) and (3.19) need not be the same as the convection velocity of the excitation, which is equal to $u = \omega/k$ in equation (3.17).

We are, in fact, following the scheme shown in Fig.1.2. The active excitation p plays the role of F_o in that figure and p_1 is part of the induced F_1, the other part being p_2. Figure 1.2 suggests that p_1 and p_2 will be computed without regard to the presence of p.

Because of the ambient velocity u_∞, and the structural motion, the induced pressure p_1 need not propagate exactly in the positive z direction. Let the direction of propagation be at an angle φ from the x-axis as shown in Fig.3.2. However, the trace of the p_1 wave along the x-axis must match that of the structural motion. Therefore,

$$k_t \cos \phi = k \tag{3.20}$$

Referring to Fig.3.2, p_1 may be expressed as

$$p_1 = P_1 e^{-i(kx+kz \tan \phi - \omega t)} \tag{3.21}$$

Substituting (3.21) into (3.18), one obtains, after simplification,

$$\cos \phi = \frac{a_1}{u - u_\infty} \tag{3.22}$$

Thus,

$$\begin{aligned} 0 < \phi \leq \frac{\pi}{2}, \quad & \text{if} \quad u \geq u_\infty \\ \frac{\pi}{2} < \phi < \pi, \quad & \text{if} \quad u < u_\infty \end{aligned} \tag{3.23}$$

and

$$k_z = k \tan \phi = \frac{k}{a_1} \sqrt{(u - u_\infty)^2 - a_1^2} \tag{3.24}$$

Only the positive square-root is retained in (3.24) since p_1 does not propagate in the negative z direction. Furthermore, no propagation is possible when

$$|u - u_\infty| < a_1$$

since k_z then becomes imaginary.

We now substitute (3.17), (3.21) and (3.24) into the boundary condition (3.19) and solve for P_1 in terms of the structural motion amplitude A,

$$p_1 = - i\rho_1 a_1 \frac{k(u - u_\infty)^2}{\sqrt{(u - u_\infty)^2 - a_1^2}} A \qquad (3.25)$$

Thus the entire p_1 field is determined. In particular, at $z = 0$,

$$(p_1)_{z=0} = - i\rho_1 a_1 \frac{k(u - u_\infty)^2}{\sqrt{(u - u_\infty)^2 - a_1^2}} A\, e^{i(\omega t - kx)} \qquad (3.26)$$

It is of interest to note that equation (3.26) can also be expressed in the form

$$(p_1)_{z=0} = - \frac{i\rho_1 a_1 \omega A \exp\left[i(\omega t - kx)\right]}{(1 + M\cos\phi)\sin\phi} \qquad (3.27)$$

where $M = u_\infty / a_1$ is the Mach number of the ambient flow.

Next we shall investigate the other induced pressure p_2 which is governed by

$$\frac{\partial^2 p_2}{\partial t^2} - a_2^2\left(\frac{\partial^2}{\partial x^2} + \frac{\partial^2}{\partial z^2}\right)p_2 = 0 \qquad (3.28)$$

and the boundary conditions:

$$\left(\frac{\partial p_2}{\partial z}\right)_{z=0} = \rho_2 \ddot{w}$$

$$\left(\frac{\partial p_2}{\partial z}\right)_{z=-d} = 0 \tag{3.29}$$

In order that the second of the above conditions can be satisfied the solution to equation (3.28) may be written as

$$p_2 = P_2 \cos k_z(z+d)e^{-i(kx-\omega t)} \tag{3.30}$$

with

$$k_z = k \tan \phi$$

$$\cos \phi = \frac{a_2}{u} \tag{3.31}$$

$$\tan \phi = \sqrt{u^2 - a_2^2}/a_2$$

obtained by letting $u_\infty = 0$ from the previous results for p_1. Of course, ϕ must now be measured clockwise from the positive x-axis. Substitute (3.30) into the first equation in (3.29). We obtain

$$P_2 = \frac{\omega^2 \rho_2}{k_z \sin k_z d} A$$

Consequently,

$$(p_2)_{z=0} = \frac{\rho_2 \omega^2 a_2 \cot k_z d}{k\sqrt{u^2 - a_2^2}} A\, e^{i(\omega t - kx)} \tag{3.32}$$

Finally, substitute (3.27) and (3.32) into (3.16). We find an equation for A:

$$A = Q^{-1} \tag{3.33}$$

where

$$Q = EIk^4 - m\omega^2 + i\rho_1 a_1 \frac{k(u - u_\infty)^2}{\sqrt{(u - u_\infty)^2 - a_1^2}} + \rho_2 a_2 \frac{ku^2 \cot k_z d}{\sqrt{u^2 - a_2^2}} \tag{3.34}$$

The required wave-number response function $\bar{H}(x,k)$ follows from

$$\bar{H}(x,k) = A(k)e^{ikx} = Q^{-1}e^{ikx} \tag{3.35}$$

Some comments about the above results are in order:

(1) The p_1 term gives rise to damping if $|u - u_\infty| > a_1$.

(2) This term should give an apparent mass effect if $|u - u_\infty| < a_1$. In such a case, the square-root $\sqrt{(u - u_\infty)^2 - a_1^2}$ should be given a negative imaginary value.

(3) For small d (shallow cavity) and $|u| > a_2$ the p_2

term gives rise to additional spring effect and for certain ranges of d value it can change to a mass effect term.

(4) When $|u| < a_2$, k_2 becomes imaginary in which case

$$\frac{\cot k_2 d}{k_z} = - \frac{\cot |k_z d|}{|k_z|}$$

Again, the p_2 term adds to the inertia of the system.

In concluding this section we remark that the above analysis has been extended to an infinite beam with periodically spaced elastic supports[10]. The solution is also mathematically exact as the one just presented herein.

IV. BUILDING RESPONSE TO GUSTY WIND

Several distinct features set the problem of wind load on buildings apart from what we have studied in the previous two sections. To understand these we refer first to the measured wind speed spectrum by Van der Hoven[11] near the earth surface, sketched in Fig.4.1.

It is generally agreed among the meteorologists that there is a spectral gap in the wind energy, although the agreement is not as universal among the fluid dynamicists. Nevertheless, we shall assume that two important humps of energy exist. The first hump accounts basically for a static load, and the second hump gives rise to the vibration of the structure. From the point of view of the usual random vibration theory, the ensemble average of a wind load must be zero. However, this interpretation has little practical significance in wind engineering, since, to do so, the important effect of static load would be entirely discounted. A more reasonable approach, therefore, is to regard the mean load to be also random.

The second distinct feature is the gradient of the mean wind as shown in Fig.4.2. In this figure z_g is called the gradient wind height. Roughly speaking, this is the height beyond which the mean wind velocity becomes nearly independent of height from the ground. It was, perhaps, Archibald[12], who first suggested a power law for the mean wind as follows:

$$\frac{u}{u_g} = \left(\frac{z}{z_g}\right)^{1/n} \qquad (4.1)$$

The parameters z_g and $1/n$ are related to another parameter z_o, called the roughness length, more frequently used by the fluid dynamicists. Davenport[13] has provided a chart where z_g and $1/n$ values can be found from the z_o value.

It is also suggested by Davenport[14] that the magnitude of the mean wind is nearly Rayleigh distributed; that is,

$$p_{|u|}(\omega) = \frac{\omega}{\sigma^2} \exp\left[-\frac{1}{2}\left(\frac{\omega}{\sigma}\right)^2\right], \qquad \omega > 0 \qquad (4.2)$$

and that the extreme value of the mean wind at a given location follows the type I extreme value distribution[15]:

$$F_{\max|u|}(\tilde{u}) = \text{Prob}\left[\max|u| \leq \tilde{u}\right]$$

$$= \exp\left\{-\exp\left[-\alpha(\tilde{u} - u_o)\right]\right\} \qquad (4.3)$$

where u_o is known as the scale parameter and α the shape parameter. From equation (4.3) one can compute, for example, the return period which is the average number of trials for $\tilde{u}$ to occur. Let

u_o = average annual maximum wind velocity

T = return period in years.

Then

$$\tilde{u} = u_o + \frac{1}{\alpha} \ln T \qquad (4.4)$$

This formula has been used frequently for flood control, the implication being that flood increases asymptotically as the logarithm of the return period, and $1/\alpha$ is the rate of increase.

We have outlined briefly the nature of the mean wind which generates a static load on a building although our main concern here is the dynamic load. Of course, the static load must be considered in the design of wind-resistant buildings. Our short remark here will serve as a reminder of its importance, before proceeding to the dynamic analysis.

In a dynamic analysis where the loading is a random process, the most important information required is, of course, the spectral density of the turbulent velocity (the second hump in Fig.4.1). Davenport[16] has suggested the following spectrum:

$$\Phi_V(\omega) = \frac{2KL^2|\omega|}{\pi^2\left[1 + \left(\frac{L\omega}{\pi u_{10}}\right)^2\right]^{4/3}} \qquad (4.3)$$

where u_{10} = mean velocity at the reference height of 10 m, L = scale of turbulence, and K = surface drag coefficient, a constant depending on the ground roughness.

Some comments about this spectrum will help to understand certain implications. First of all, it is assumed that the scale of turbulence L is independent of height. Berman[17] has reached a similar conclusion although intuition would suggest that L grows with height. Secondly, in equation (4.3), the spectrum tends to zero linearly at small ω, and tends to zero as $\omega^{-5/3}$ for large ω. According to von Karman, the spectrum of a typical turbulence does vary as $\omega^{-5/3}$ at high frequencies but it should be rather flat (a zero-power law) at low frequencies. Therefore, equation (4.3) is not in agreement with the von Karman spectrum at low frequencies. In what follows we shall not subscribe to a specific spectrum. We merely assume that some suitable spectrum will be used in actual computations.

Again, we shall choose a simple model shown in Fig.4.3, a box type structure having two degrees of freedom, recently studied by Vaicaitis, Shinozuka, and Takeno[18]. It is assumed that the structure moves only in the x_1 and x_2 directions, and the mean wind is directed in the x_1 direction. (A general mean wind direction would add such complexities as the more complex vortex shedding patterns, etc.). We further assume that the forces generated by the wind depend on the velocity of the wind relative to the structure, and that these forces can be represented by a drag D in the direction of the relative velocity and a lift L perpendicular to the relative velocity.

Then, referring to Fig.4.4, the relative velocity is

$$V_r = \left\{(u + v_1 - \dot{w}_1)^2 + (v_2 - \dot{w}_2)^2 + v_3^2\right\}^{1/2} \qquad (4.4)$$

Following Etkin[19] the drag force D may be expressed as

$$D = \frac{1}{2}\,\rho\, C_D\, A\, V_r^2 + \frac{\pi}{4}\,\rho\, C_M\, A\, b\, \dot{V}_r \qquad (4.5)$$

where ρ = air density, A = windward area of the structure, b = lateral dimension of the structure and C_D and C_M are constants. Strictly speaking, equation (4.5) is obtained from the strip theory of aerodynamics, but we shall assume that it remains valid for the present case with suitable adjustments in the values of C_D and C_M.

The lift L is due to vortex shedding at the corners of the structure. Under an ideal smooth flow condition, a vortex shedding frequency can be identified, and the lift varies sinusoidally at a frequency ω_s, related to a Strouhal number

$$s = \frac{\omega_s b}{u} \tag{4.6}$$

In a turbulent flow the situation is much more complicated. However, we shall assume that the following representation is adequate

$$L = \frac{1}{2} \rho A C_L V_r^2 \cos\left[\omega_s t + \psi(t)\right] \tag{4.7}$$

where $\psi(t)$ is a slowly varying random process.

The components of D and L in the x_1 and x_2 directions are given by (referred to Fig.4.4)

$$\begin{aligned} D_1 &= D \cos \alpha_1 \\ D_2 &= D \sin \alpha_1 \cos \alpha_2 \\ L_1 &= L \sin \alpha_1 \cos \alpha_2 \\ L_2 &= L \cos \alpha_1 \end{aligned} \tag{4.8}$$

For convenience, we shall use the symbols

$$g_1 = \frac{1}{2} \rho C_D A$$

$$g_2 = \frac{\pi}{4} \rho C_M Ab \qquad (4.9)$$

$$g_3 = \frac{1}{2} \rho C_L A$$

and

$$X_i = \frac{V_i - \dot{W}_i}{u}, \text{ with } \dot{W}_3 = 0 \qquad (4.10)$$

Then one finds

$$V_r = u\{(1 + X_1)^2 + X_2^2 + X_3^2\}^{1/2} \qquad (4.11)$$

$$\dot{V}_r = u\{(1 + X_1)\dot{X}_1 + X_2\dot{X}_2 + X_3\dot{X}_3\}\{(1 + X_1)^2 + X_2^2 + X_3^2\}^{-1/2} \qquad (4.12)$$

Introducing the following reasonable assumption for the orders of magnitude of various quantities involved[18]

$$|X_1| \gg |X_2| \gg |X_3|, \qquad |\dot{X}_1| \gg |\dot{X}_2| \gg |\dot{X}_3|$$

one obtains the equations of motion in the forms of

$$\ddot{W}_1 + 2\bar{\zeta}_1\bar{\omega}_1\dot{W}_1 + \bar{\omega}_1^2 W_1 = D_1/m \qquad (4.13)$$

$$\ddot{W}_2 + 2\zeta_2\omega_2\dot{W}_2 + \omega_2^2 W_2 = L_2/m \qquad (4.14)$$

where

$$D_1 = g_1(u^2 + V_1^2 + \dot{W}_1^2 + 2uV_1 - 2u\dot{W}_1 - 2V_1\dot{W}_1) + g_2(\dot{V}_1 - \ddot{W}_1)$$

$$L_2 = g_3\left[(u + V_1 - \dot{W}_1)^2 + \frac{1}{2}(V_2 - \dot{W}_2)^2\right]\cos\left[\omega_s t + \psi(t)\right] \quad (4.15)$$

Substituting (4.15) into equations (4.13) and (4.14), and introducing

$$Y_1 = W_1 - \frac{g_1(u^2 + E\left[V_1^2\right])}{m\,\bar{\omega}_1^2}$$

$$Y_2 = W_2$$

$$\omega_1^2 = \bar{\omega}_1^2\left(\frac{m}{m + g_2}\right)$$

$$\zeta_1 = \left(\bar{\zeta}_1 + \frac{ug_1}{m\bar{\omega}_1}\right)\sqrt{\frac{m}{m + g_2}}$$

$$\beta = \frac{g_1}{m + g_2}, \qquad \gamma = g_3/2m$$

the equations of motion are simplified to

$$\ddot{Y}_1 + 2\zeta_1\omega_1\dot{Y}_1 + 2\beta V_1\dot{Y}_1 - \beta\,\dot{Y}_1^2 + \omega_1^2 Y_1 = F_1(t) \quad (4.16)$$

$$\ddot{Y}_2 + 2\zeta_2\omega_2\dot{Y}_2 + 2\gamma V_2\cos\left[\omega_s t + \psi(t)\right]\dot{Y}_2$$

$$- \gamma\cos\left[\omega_s t + \psi(t)\right]\dot{Y}_2^2 + \omega_2^2 Y_2$$

$$= 2\gamma\left[(u + V_1 - \dot{Y}_1)^2 + \frac{1}{2}V_2^2\right]\cos\left[\omega_s t + \psi(t)\right] \quad (4.17)$$

where

$$F_1(t) = \beta\{V_1^2 - E\left[V_1^2\right] + 2uV_1 + \frac{g_2}{g_1}\dot{V}_1\} \qquad (4.18)$$

Some comments are in order. Equations (4.16) and (4.17) are nonlinear equations with Y_1 independent of Y_2, but Y_2 dependent of Y_1. It is generally agreed that the turbulent velocity processes V_1 and V_2 are Gaussian; therefore, the forcing functions on the righthand sides of (4.16) and (4.17) are non-Gaussian. Both equations involve parametric excitations which can cause instability of the system.

We shall consider the simpler equation, (4.16) first. Since the excitations, V_1 on the lefthand side and F_1 on the righthand side (related to V_1) are broad-band processes, the method of stochastic averaging of Stratonovich[20] can be applied if the system damping is low. Stratonovich's method has been proved rigorously by Khasminskii[21] who showed that under suitable conditions* the response tends weakly to Markovian governed by an equivalent Itô equation.

Since $Y_1(t)$ is a narrow-band process we can write

$$Y_1 = A(t)\cos\Gamma, \qquad \dot{Y}_1 = -\omega_1 A(t)\sin\Gamma, \quad \Gamma = \omega_1 t + \theta(t) \qquad (4.19)$$

* See Appendix I

where A(t) and O(t) are slowly varying random processes. Equation (4.16) can then be substituted by two first order equations

$$\dot{A} = \frac{1}{\omega_1}\left[-2\zeta_1\omega_1^2 A \sin\Gamma - 2\beta V_1\omega_1 A \sin\Gamma - \beta\omega_1^2 A \sin^2\Gamma - F_1(t)\right]\sin\Gamma \qquad (4.20)$$

$$\dot{\theta} = \frac{1}{\omega_1 A}\left[\text{same expression}\right]\cos\Gamma$$

According to Bogolibov and Mitropolskii[22] since A and O vary only slowly, (4.20) may be approximated by a set of averaged equations where the right hand sides are replaced by the long-time average values. The net effect of this averaging process is that all "oscillatory terms" can be dropped, resulting in

$$\begin{aligned}\dot{A} &= -\zeta_1\omega_1 A + G_1\\ \dot{\theta} &= G_2\end{aligned} \qquad (4.21)$$

where G_1 and G_2 are the components of a row vector $\underline{G}$ whose transpose is

$$\underline{G}' = \begin{Bmatrix} G_1 \\ G_2 \end{Bmatrix} = \begin{Bmatrix} -2\beta A\, V_1 \sin^2\Gamma - \frac{1}{\omega_1} F_1 \sin\Gamma \\ -2\beta V_1 \sin\Gamma\cos\Gamma - \frac{1}{\omega_1 A} F_1 \cos\Gamma \end{Bmatrix} \qquad (4.22)$$

Terms of the type of $V_1 \sin^2 \Gamma$, $F_1 \sin \Gamma$, etc. are called non-oscillatory terms which must be treated by the method of stochastic averaging (or Stratonovich-Khasminskii limit theorem). The S-K limit theorem asserts that $\underline{Z} = \{A, \theta\}$ approaches weakly to a Markov vector and, in the limit, is governed by the Itô equation

$$d\underline{Z} = \underline{m}\, dt + \underline{\sigma}\, d\underline{B} \tag{4.23}$$

where $\underline{B}$ is a vector of unit Brownian (Wiener) processes, and $\underline{m}$ and $\underline{\sigma}$ are computed from

$$\underline{m} = \begin{Bmatrix} -\zeta_1 \omega_1 A \\ 0 \end{Bmatrix} + \int_{-\infty}^{0} \left\langle E\left[\left(\frac{\partial G}{\partial z}\right)'_t (G)_{t+\tau} \right] \right\rangle d\tau$$

$$\underline{\sigma}\, \underline{\sigma}' = \int_{-\infty}^{\infty} \left\langle E\left[(G')_t\, (G)_{t+\tau} \right] \right\rangle d\tau \tag{4.24}$$

in which $\langle\ \rangle$ denotes a time average and the subscript t or $t+\tau$ indicates that the quantity is to be evaluated at t or $t+\tau$. Substituting (4.22) into (4.24), we obtain, after rather long algebraic work,

$$m_1 = \left[-\zeta_1 \omega_1 + \pi\beta^2\, \phi_{V_1}(0) + \frac{3}{2}\, \pi\beta^2\, \phi_{V_1}(2\omega_1) \right] A + \frac{\pi}{2\omega_1^2 A}\, \phi_{F_1}(\omega)$$

$$m_2 = 0 \tag{4.25}$$

$$(\sigma\,\sigma')_{11} = \beta^2 A^2 (2\pi)\left[\phi_{V_1}(0) + \frac{1}{2}\,\phi_{V_1}(2\omega_1)\right] + \frac{\pi}{\omega_1^2}\,\phi_{F_1}(\omega_1)$$

$$(\sigma\,\sigma')_{22} = \pi\left[\beta^2\,\phi_{V_1}(2\omega_1) + \frac{1}{\omega_1^2 A^2}\,\phi_{F_1}(\omega_1)\right] \qquad (4.26)$$

The non-diagonal terms of the $\underline{\sigma}\ \underline{\sigma}'$ matrix turn out to be zero. The spectral density of F_1 must, of course, be related to the spectral density of V_1. It can be shown by use of (4.18) that

$$\phi_{F_1}(\omega) = (2u\beta)^2\,\phi_{V_1}(\omega) + (g_2/g_1)^2\beta^2\omega^2\,\phi_{V_1}(\omega)$$

$$+\ 2\beta^2 \int_{-\infty}^{\infty} \phi_{V_1}(\omega-\omega')\,\phi_{V_1}(\omega')\,d\omega' \qquad (4.27)$$

It is of interest to note that in the equivalent Itô equation for the system, (4.23), A turns out to be independent of θ. This means that although (A,θ) constitutes a weak Markov vector, A is a weak Markov (scalar) process itself. Generally the component of a Markov vector is not necessarily a Markov process.

The Fokker-Planck equation for the transition probability density of the Markov vector (A,O) follows immediately from (4.25) and (4.26). Denote this transition probability density by $q = q(A,\theta,t\,|\,A_o,\theta_o,t_o)$.

$$\frac{\partial q}{\partial t} = -\frac{\partial}{\partial A}(m_1 q) + \frac{1}{2}\frac{\partial^2}{\partial A^2}\left[(\sigma\,\sigma')_{11} q\right] + \frac{1}{2}\frac{\partial^2}{\partial \theta^2}\left[(\sigma\,\sigma')_{22} q\right] \qquad (4.28)$$

Substituting (4.25) and (4.26) into (4.28):

$$\frac{\partial q}{\partial t} = (a_2 A^2 + e)\frac{\partial^2 q}{\partial A^2} + (a_1 A - \frac{e}{A})\frac{\partial q}{\partial A} + (a_o + \frac{e}{A^2})q + $$

$$+ (b_o + \frac{e}{A^2})\frac{\partial^2 q}{\partial \theta^2} \qquad (4.29)$$

where

$$a_2 = \frac{\pi}{2}\left[2\beta^2\ \phi_{V_1}(0) + \beta^2\ \phi_{V_1}(2\omega_1)\right]$$

$$a_1 = \zeta_1\omega_1 + 3\pi\ \beta^2\ \phi_{V_1}(0) + \frac{1}{2}\ \pi\beta^2\ \phi_{V_1}(2\omega_1)$$

$$a_o = \zeta_1\omega_1 + \pi\beta^2\ \phi_{V_1}(0) - \frac{1}{2}\ \pi\beta^2\ \phi_{V_1}(2\omega_1) \qquad (4.30)$$

$$b_o = \frac{\pi}{2}\ \beta^2\ \phi_{V_1}(2\omega_1)$$

$$e = \frac{\pi}{2\omega_1^2}\ \phi_{F_1}(\omega_1)$$

In equation (4.29) the ranges for A and θ are $0 \leq A < \infty$ and $-\infty < \theta < \infty$, respectively. Therefore, the Fokker-Planck equation for A alone can be obtained by integrating (4.29) over Θ, since a probability density and its derivatives must vanish at infinity. Let $q_A = q_A(A,t|A_o,t_o)$.

$$\frac{\partial q_A}{\partial t} = (a_2 A^2 + e)\frac{\partial^2 q_A}{\partial A^2} + (a_1 A - \frac{e}{A})\frac{\partial q_A}{\partial A}$$

$$+ (a_o + \frac{e}{A^2})\ q_A \qquad (4.31)$$

This equation could have been obtained by use of the first row of equation (4.23).

A Fokker-Planck equation for θ alone cannot be obtained by integrating (4.29) over A, however, since $0 \leq A < \infty$, and q and its derivatives need not vanish at $A = 0$.

Solution for either (4.29) or (4.31) appears to be difficult. One then wonders if it is possible to obtain a stationary solution for (4.31) by setting $\partial q_A/\partial t$ equal to zero. For this purpose, it is simpler to return to (4.28), since such a stationary solution, if it exists, would be governed by

$$- m_1 p_A + \frac{1}{2}\frac{\partial}{\partial A}\left[(\sigma\,\sigma')_{11}\, p_A\right] = 0 \qquad (4.32)$$

Equation (4.32) can be solved when

$$\zeta_1\omega_1 > \pi\,\beta^2\,\phi_{V_1}(2\omega_1)$$

However, this solution will not be discussed here.

For the time being, we shall restrict ourselves to some statistical properties of the response. From the first row of equation (4.23):

$$dA = \left\{\left[-\zeta\omega_1 + \pi\beta^2\,\phi_{V_1}(0) + \frac{3}{2}\,\pi\beta^2\,\phi_{V_1}(2\omega_1)\right] A + \frac{\pi}{2\omega_1^2 A}\phi_{F_1}(\omega_1)\right\} dt$$

$$+ \left\{\pi\beta^2\left[2\phi_{V_1}(0) + \phi_{V_1}(2\omega_1)\right] A^2 + \frac{\pi}{\omega_1^2}\,\phi_{F_1}(\omega)\right\}^{\frac{1}{2}} dB_1 \qquad (4.33)$$

From this we can find an expression for the differential dA^2 using Ito's differential rule[23]:

$$dA^2 = \{2[-\zeta_1\omega_1 + 2\pi\beta^2 \phi_{V_1}(0) + 2\pi\beta^2 \phi_{V_1}(2\omega_1)] A^2 + \frac{2\pi}{\omega_1^2} \phi_{F_1}(\omega_1)\} dt + 2A\sqrt{(\sigma\sigma')_{11}}\, dB_1$$

Taking the ensemble average

$$d(E[A^2]) = \{2[-\zeta_1\omega_1 + 2\pi\beta^2 \phi_{V_1}(0) + 2\pi\beta^2 \phi_{V_1}(2\omega_1)]E[A^2] + \frac{2\pi}{\omega_1^2} \phi_{F_1}(\omega_1)\} dt \quad (4.34)$$

In obtaining (4.34) we have used the fact that the response at t is independent of dB(t). This is related to the Markov property, the basis for the Itô calculus. Equation (4.34) is solved readily:

$$E[A^2] = (A_o^2 - A_s^2)\, e^{-2\delta(t-t_o)} + A_s^2 \quad (4.35)$$

where $A_o = A(t_o)$ = initial state, and $A_s^2 = E[A^2]$ at the steady state, and

$$\delta = \zeta_1\omega_1 - 2\pi\beta^2 \phi_{V_1}(0) - 2\pi\beta^2 \phi_{V_1}(2\omega_1) \quad (4.36)$$

It is seen that the second moment instability can occur when

$$\zeta_1\omega_1 < 2\pi\beta^2 \phi_{V_1}(0) + 2\pi\beta^2 \phi_{V_1}(2\omega_1) \quad (4.37)$$

When the system is stable, the steady state amplitude is given by

$$A_s^2 = \frac{\pi}{\omega_1^2}\left[\zeta_1\omega_1 - 2\pi\beta^2\,\Phi_{V_1}(0) - 2\pi\beta^2\,\Phi_{V_1}(2\omega_1)\right]^{-1}\Phi_{F_1}(\omega_1) \tag{4.38}$$

It is interesting to note that the stability in mean square is related to the spectral density of V_1 evaluated at twice the natural frequency, a fact well-known in the stability study of deterministic systems. Ariaratnam[24] also found such a dependence in moment stabilities of linear stochastic systems. We have found in addition, that the spectral density of the parametric excitation at zero frequency is also significant if the excitation occurs in the damping parameter. The negative damping term associated with $\dot{Y}^2$ in equation (4.16) has no effect on stability, contrary to the speculation of some authors. This term was dropped when the oscillatory terms were eliminated, going from (4.20) to (4.21). As long as the motion of the system is nearly cyclic, a damping term proportional to the square of the velocity will supply energy to the system in half a cycle while it will absorb about the same amount of energy in the other half of the cycle. Therefore, as far as stability is concerned the Y_1 motion behaves linearly. The forcing function F_1 does not contribute to moment instability *per se*.

Recognizing the basic stability behaviour of the Y_1 motion we can apply Infante's result[25] for almost sure (sample) stability for linear systems in this case. Thus, a sufficient condition for sample stability for equation (4.16) is given by

$$E\left[V_1^2\right] < \frac{\zeta^2\omega_1^2}{\beta^2(\zeta^2 + 1)} \tag{4.39}$$

Infante's result is valid for any V_1 process. Kozin and Wu[26] have shown that considerably sharper stability boundary can be stablished by using the information that V_1 is Gaussian.

We now turn our attention to equation (4.17) governing the motion Y_2. Analysis of this equation is much more difficult and the materials to be presented below are tentative. However, they may serve as a basis for discussion. We note that the excitations involve $\dot{Y}_1$ which is a narrow-band process, and V_1 and V_2 which are broad-band processes. All of these are multiplied by a narrow-band process, $\cos\left[\omega_s t + \Psi\right]$.

To simplify equation (4.17) somewhat, we shall calculate first the spectral density of $F_2(t)$ where

$$F_2(t) = (u + V_1 - \dot{Y}_1)^2 + \frac{1}{2} V_2^2 \tag{4.40}$$

The correlation function of $u + V_1 - \dot{Y}_1$, denoted by $R_1(\tau)$, is

$$R_1(\tau) = u^2 + R_{V_1}(\tau) + \omega_1 E\{V_1(t)A(t+\tau)\sin(\omega_1 t + \omega_1\tau + \theta_{t+\tau})$$

$$+ V_1(t+\tau)A(t)\sin(\omega_1 t + \theta_t)\} \qquad (4.41)$$

$$+ \omega_1^2 E\{A(t)\sin(\omega_1 t + \theta_t)A(t+\tau)\sin(\omega_1 t + \omega_1\tau + \theta_{t+\tau})\}$$

The two unevaluated expectations in equation (4.41) cannot be calculated exactly, but it seems reasonable to approximate their values by use of the stochastic averaging procedure similar to that used in equation (4.24). Then the second expected value is approximately

$$\frac{1}{2} A_s^2 \cos \omega_1\tau$$

and the first expected value is approximately zero since, for example,

$$E\{V_1(t)A(t+\tau)\sin(\omega_1 t + \omega_1\tau + \theta_{t+\tau})\}$$

$$\approx \int_{-\infty}^{0} \left\langle E\left[V_1(t)A(t+\tau)\left(\frac{\partial}{\partial\theta}\sin\Gamma\right)_t \left(\frac{\partial G_2}{\partial V_1}\right)_{t+\tau} V_1(t+\tau)\right]\right\rangle d\tau$$

$$\approx \int_{-\infty}^{0} \left\langle A\, R_{V_1}(\tau)\cos\Gamma_t(-\beta\sin 2\Gamma)_{t+\tau}\right\rangle d\tau \approx 0$$

where we have used the second row of (4.22) to evaluate $(\partial G_2/\partial V_1)$ which is the change in θ due to the variation of V_1. Therefore, we obtain an approximation

$$R_1(\tau) \approx u^2 + R_{V_1}(\tau) + \frac{\omega_1^2}{2} A_s^2 \cos \omega_1 \tau \qquad (4.42)$$

We have assumed that the Y_1 motion is stable so that its stationary meansquare amplitude A_s^2 exists.

What we really need is, of course, the correlation function of $(u + V_1 - \dot{Y}_1)^2$. Again, we face a difficulty since $\dot{Y}_1$ is not Gaussian although V_1 is Gaussian. If, however, V_1 dominates over $\dot{Y}_1$ (a reasonable assumption) then $u + V_1 - \dot{Y}_1$ is approximately Gaussian. On this basis, we can write

$$R_{(u+V_1-\dot{Y}_1)^2} \approx 2R_1^2(\tau) + R_1^2(0) - 2u^4 \qquad (4.43)$$

The correlation function for $\frac{1}{2} V_2^2$ can easily be obtained. Assuming V_2 to be Gaussian, we have

$$R_2(\tau) = \frac{1}{4} \left[2R_{V_2}^2(\tau) + R_{V_2}^2(0) \right] \qquad (4.44)$$

To compute the cross-correlation between $(u + V_1 - \dot{Y}_1)^2$ and $\frac{1}{2} V_2^2$, we neglect the correlation between $\dot{Y}_1$ and V_2 arriving at

$$R_3(\tau) = \frac{1}{2} u^2 R_{V_2}(0) + \frac{1}{2} R_{V_2}(0)\, R_{V_1}(0) + \frac{\omega_1^2}{4} A_s^2 R_{V_2}(0) + \frac{1}{2}\left[R_{V_1V_2}(\tau) + R_{V_2V_1}(\tau) \right] \qquad (4.45)$$

Finally, the correlation function for $F_2(t)$ is obtained by summing up (4.43), (4.44) and (4.45):

$$R_{F_2}(\tau) = J_1 + 2J_2 \cos \omega_1\tau + 2J_3 \cos 2\omega_1\tau$$

$$+ J_4R_{V_1}(\tau) + J_5R^2_{V_1}(\tau) + J_6R^2_{V_2}(\tau)$$

$$+ \frac{1}{2} J_7 \left[R_{V_1V_2}(\tau) + R_{V_2V_1}(\tau)\right]$$

$$+ 2J_8R_{V_1}(\tau) \cos \omega_1\tau \qquad (4.46)$$

The spectral density of $F_2(t)$ is the Fourier transform of (4.46); thus

$$\Phi_{F_2}(\omega) = J_1\delta(\omega) + J_2\left[\delta(\omega-\omega_1) + \delta(\omega+\omega_1)\right]$$

$$+ J_3\left[\delta(\omega-2\omega_1) + \delta(\omega+2\omega_1)\right]$$

$$+ J_4\Phi_{V_1}(\omega) + J_5 \int_{-\infty}^{\infty} \Phi_{V_1}(\omega')\Phi_{V_1}(\omega-\omega')d\omega'$$

$$+ J_6 \int_{-\infty}^{\infty} \Phi_{V_2}(\omega')\Phi_{V_2}(\omega-\omega')d\omega'$$

$$+ J_7\mathrm{Re}\left[\Phi_{V_1V_2}(\omega)\right] +$$

$$+ J_8\left[\Phi_{V_1}(\omega-\omega_1) + \Phi_{V_1}(\omega+\omega_1)\right] \qquad (4.47)$$

In equations (4.46) and (4.47),

$$J_1 = \{u^2 + R_{V_1}(0) + (\omega_1^2/2)A_s^2\}^2 + \frac{1}{4} R_{V_2}^2(0)$$

$$+ \frac{1}{2} u^2 R_{V_2}(0) + \frac{1}{2} R_{V_2}(0)\, R_{V_1}(0)$$

$$+ \frac{\omega_1^2}{4} A_s^2\, R_{V_2}(0) + \frac{\omega_1^4}{4} A_s^4$$

$$J_2 = u^2\omega_1^2 A_s^2, \qquad J_3 = \frac{1}{8}\,\omega_1^4 A_s^4$$

$$J_4 = 4u^2, \qquad J_5 = 2, \qquad J_6 = \frac{1}{2},$$

$$J_7 = 1, \qquad J_8 = \frac{1}{4}\,\omega_1^2 A_s^2$$

To make the remaining analysis tractable, we shall replace $F_2(t)$ by the sum of the following forces:

(1) A static force $Z_1 = \sqrt{J_1}$

(2) A sinusoidal random force $Z_2 \cos(\omega_1 t + \alpha_2)$ with

$$E\left[Z_2^2\right] = 4J_2.$$

(3) A sinusoidal random force $Z_3 \cos(2\omega_1 t + \alpha_3)$ with

$$E\left[Z_3^2\right] = 4J_3.$$

(4) A broad-band noise $\xi(t)$ with a spectrum equal to the sum of J_4 to J_8 terms in (4.47).

We shall assume that the forces (2) through (4) be statistically independent of each other, and independent of the coefficient process cos $(\omega_s t + \psi)$. With this substitution, equation (4.17) now becomes

$$\ddot{Y}_2 + 2\zeta_2\omega_2\dot{Y}_2 + (2\gamma\, V_2\dot{Y}_2 - \gamma\dot{Y}_2^2)\cos\left[\omega_s t + \psi(t)\right]$$

$$+ \omega_2^2 Y_2 = 2\gamma\left[Z_1 + Z_2\cos\,(\omega_1 t + \alpha_2) + Z_3\cos\,(2\omega_1 t + \alpha_3)\right.$$

$$\left. + \xi(t)\right]\cos\,(\omega_s t + \psi) \qquad (4.48)$$

We digress to investigate the nature of various terms in (4.48). We note, first, that the product of a broad-band noise and a narrow-band noise, such as $V_2\cos\,(\omega_s t + \psi)$ or $\xi(t)\cos\,(\omega_s t + \psi)$, is a broad-band noise, provided that the band-width of $V_2(t)$ or $\xi(t)$ extends beyond ω_s. It can be shown that the effect of the narrow-band factor $\cos\,(\omega_s t+\psi)$ is to reduce to one-half the height of the original broad-band spectrum. Therefore, $V_2\cos\,(\omega_s t + \psi)$ and $\xi(t)\cos\,(\omega_s t+\psi)$ in equation (4.48) may be replaced, respectively, by $\frac{1}{\sqrt{2}}\,V_2$ and $\frac{1}{\sqrt{2}}\,\xi(t)$. The products of cosine functions in equation (4.48) also may be replaced by the sums of such functions. Recognizing these, equation (4.48) may be rewritten in a more manageable form as follows:

$$\ddot{Y}_2 + 2\zeta_2\omega_2\dot{Y}_2 + \sqrt{2}\,\gamma\,V_2\dot{Y}_2 - \gamma\cos\left[\omega_s t + \psi\right]\dot{Y}_2^2$$

$$+ \omega_2^2 Y_2 = 2\gamma\,Z_1\cos(\omega_s t+\psi) + \gamma\,Z_2\{\cos\left[(\omega_1+\omega_s)t + (\alpha_2+\psi)\right]$$

$$+ \cos\left[(\omega_1-\omega_s)t + (\alpha_2-\psi)\right]\} + \gamma\,Z_3\{\cos\left[(2\omega_1+\omega_s)t + (\alpha_3+\psi)\right]$$

$$+ \cos\left[(2\omega_1-\omega_s)t + (\alpha_3-\psi)\right]\} + \sqrt{2}\,\gamma\,\xi\,t) \qquad (4.49)$$

On the left hand side of this equation we find a broad-band parametric excitation V_2 associated with the velocity $\dot{Y}_2$, and a narrow-band parametric excitation associated with the squared velocity $\dot{Y}_2^2$. The non-parametric excitation terms are grouped on the right hand side of the equation, including five narrow-band terms and one broadband term.

Equation (4.49) still appears extremely complicated and a solution for the general case cannot be obtained at the present time. However, we can investigate the most important case where the response is a narrow-band random process with a peak frequency ω. Under this assumption we may write

$$Y_2 = \tilde{X}_1\cos\omega t + \tilde{X}_2\sin\omega t$$
$$\dot{Y}_2 = -\tilde{X}_1\omega\sin\omega t + \tilde{X}_2\omega\cos\omega t \qquad (4.50)$$

where X_1 and X_2 are slowly varying random processes. Then

(4.49) may be replaced by a pair of first order equations as follows:

$$\dot{\tilde{X}}_1 = -\omega(\tfrac{1}{2}\tilde{X}_1 \sin 2\omega t + \tilde{X}_2 \sin^2\omega t) + 2\zeta_2\omega_2 \times$$

$$(-\tilde{X}_1 \sin^2\omega t + \tfrac{1}{2}\tilde{X}_2 \sin 2\omega t) + \sqrt{2}\,\gamma\, V_2 \times$$

$$(-\tilde{X}_1 \sin^2\omega t + \tfrac{1}{2}\tilde{X}_2 \sin 2\omega t) - \gamma\omega \cos(\omega_s t + \psi) \times$$

$$\sin \omega t(-\tilde{X}_1 \sin \omega t + \tilde{X}_2 \cos \omega t)^2 + \frac{\omega_2^2}{\omega} \times$$

$$(\tfrac{1}{2}\tilde{X}_1 \sin 2\omega t + \tilde{X}_2 \sin^2\omega t) - \frac{1}{\omega} \sin \omega t \times$$

(r.h.s. of equation (4.49))

$$\dot{\tilde{X}}_2 = \omega(\tilde{X}_1 \cos^2\omega t + \tfrac{1}{2}\tilde{X}_2 \sin 2\omega t) + 2\zeta_2\omega_2(\tfrac{1}{2} X_1 \sin 2\omega t$$

$$- X_2 \cos^2\omega t) - \sqrt{2}\,\gamma\, V_2(-\tfrac{1}{2}\tilde{X}_1 \sin 2\omega t + \tilde{X}_2 \cos^2\omega t)$$

$$+ \gamma\omega \cos(\omega_s t + \psi)\cos \omega t(-\tilde{X}_1 \sin \omega t + \tilde{X}_2 \cos \omega t)^2$$

$$- \frac{\omega_2^2}{\omega}(\tilde{X}_1 \cos^2\omega t + \tfrac{1}{2}\tilde{X}_2 \sin 2\omega t) + \frac{1}{\omega} \cos \omega t \times$$

(r.h.s. of equation (4.49)) (4.51)

If the relaxation time of the $\tilde{X}_1$ and $\tilde{X}_2$ processes is much larger than the correlation times of both $V_2(t)$ and $\xi(t)$, we expect that $(\tilde{X}_1, \tilde{X}_2)$ approaches weakly to a Markov vector governed by an Itô equation of the form of (4.23). To obtain this Itô equation we again use time-averaging to eliminate the time variable for terms not involving $V_2(t)$ nor $\xi(t)$, and use stochastic averaging for terms involving $V_2(t)$ or $\xi(t)$. However, since the value for ω has not been specified, the time-averaging will be performed in two steps. In the first step, terms that can be time-averaged without knowing the exact ω value will be dealt with. This leads to

$$\dot{\tilde{X}}_1 = -\left(\frac{\omega^2 - \omega_2^2}{2\omega}\right)\tilde{X}_2 - \zeta_2\omega_2\tilde{X}_1 + \sqrt{2}\,\gamma\, V_2(-\tilde{X}_1 \sin^2\omega t$$

$$+ \frac{1}{2}\tilde{X}_2 \sin 2\omega t) - \frac{\gamma\omega}{2}\cos(\omega_s t + \psi)\sin\omega t \times$$

$$\left[(\tilde{X}_1^2 + \tilde{X}_2^2) - (\tilde{X}_1^2 - X_2^2)\cos 2\omega t - 2\tilde{X}_1\tilde{X}_2 \sin 2\omega t\right]$$

$$- \frac{\sin\omega t}{\omega}\,(\text{r.h.s. of equation } (4.49))$$

$$\dot{\tilde{X}}_2 = \frac{\omega^2 - \omega_2^2}{2\omega} \tilde{X}_1 - \zeta_2 \omega_2 \tilde{X}_2 - \sqrt{2}\, \gamma\, V_2(- \tfrac{1}{2} \tilde{X}_1 \sin 2\omega t$$

$$+ \tilde{X}_2 \cos^2 \omega t) + \frac{\gamma\omega}{2} \cos (\omega_s t + \psi) \cos \omega t \ \times$$

$$\left[(\tilde{X}_1^2 + \tilde{X}_2^2) - (\tilde{X}_1^2 - \tilde{X}_2^2) \cos 2\omega t - 2\tilde{X}_1 \tilde{X}_2 \sin 2\omega t\right]$$

$$+ \frac{\cos \omega t}{\omega} \text{ (r.h.s. of equation (4.49)).} \qquad (4.52)$$

These equations are valid for any ω sufficiently close* to ω_2.

The second step of time-averaging requires that the value for ω be specified. We shall consider the possible "worst" case where $\omega_s \approx \omega_2$. On rare occasions, ω_s, ω_2 and one of the narrow-band frequencies associated with Z_2 or Z_3 may be all very close to one another, but we shall not consider such a case here.

Setting $\omega = \omega_s$ in (4.52) and carrying out time-average for the remaining terms not involving $V_2(t)$ nor $\xi(t)$, we obtain

* For the averaged equations to be valid all the terms on the right hand side must be small in some sense. See, also Appendix I.

$$\dot{\tilde{X}}_1 = -\left(\frac{\omega_s^2 - \omega_2^2}{2\omega_s}\right)\tilde{X}_2 - \zeta_2\omega_2\tilde{X}_1 + \frac{\gamma\omega_s}{8}\left[(3\tilde{X}_1^2 + \tilde{X}_2^2)\sin\Psi\right.$$

$$\left. + 2\tilde{X}_1\tilde{X}_2\cos\Psi\right] + \frac{\gamma}{\omega_s} Z_1 \sin\Psi$$

$$+ \sqrt{2}\,\gamma\, V_2(-\tilde{X}_1 \sin^2\omega_s t + \frac{1}{2}\tilde{X}_2 \sin 2\omega_s t)$$

$$- \frac{1}{\omega_s}\sqrt{2}\,\gamma\,(\sin\omega_s t)\xi(t)$$

$$\dot{\tilde{X}}_2 = \left(\frac{\omega_s^2 - \omega_2^2}{2\omega_s}\right)\tilde{X}_1 - \zeta_2\omega_2\tilde{X}_2 +$$

$$\frac{\gamma\omega_s}{8}\left[(\tilde{X}_1^2 + 3\tilde{X}_2^2)\cos\Psi + 2\tilde{X}_1\tilde{X}_2\sin\Psi\right] + \frac{\gamma}{\omega_s} Z_1 \cos\Psi$$

$$- \sqrt{2}\,\gamma\, V_2(-\frac{1}{2}\tilde{X}_1 \sin 2\omega_s t + \tilde{X}_2 \cos^2\omega_s t)$$

$$+ \frac{1}{\omega_s}\sqrt{2}\,\gamma\,(\cos\omega_s t)\xi(t). \tag{4.53}$$

To obtain an Itô equation corresponding to (4.53) we use equation (4.24) with obvious modifications to suit the present case. In particular, the first term in the <u>m</u> expression is replaced by a column whose elements are the right hand sides of the expressions in (4.53) after deleting $V_2(t)$ and $\xi(t)$ terms, and G_1 and G_2 are

$$G_1 = \sqrt{2}\ \gamma\ V_2(-\tilde{X}_1\ \sin^2\omega_s t + \frac{1}{2}\tilde{X}_2\ \sin 2\omega_s t)$$

$$- \frac{1}{\omega_s}\ \sqrt{2}\ \gamma\ (\sin\ \omega_s t)\xi(t)$$

$$G_2 = -\ \sqrt{2}\ \gamma\ V_2(-\ \frac{1}{2}\tilde{X}_1\ \sin 2\omega t + \tilde{X}_2\ \cos^2\omega t)$$

$$+ \frac{1}{\omega_s}\ \sqrt{2}\ \gamma\ (\cos\ \omega_s t)\xi(t). \qquad (4.54)$$

Then application of (4.24) yields

$$m_1 = -\ (\frac{\omega_s^2 - \omega_2^2}{2\omega_s})\tilde{X}_2 - \zeta_2\omega_2\tilde{X}_1 + \frac{\gamma\omega_s}{8}\Big[(3\tilde{X}_1^2 + X_2^2)\sin\ \Psi$$

$$+ 2\tilde{X}_1\tilde{X}_2\ \cos\ \Psi\Big] + \frac{\gamma}{\omega_s}\ Z_1\ \sin\ \Psi + \frac{\pi\gamma^2}{2}\ \tilde{X}_1\Big[\Phi_{V_2}(0) + \Phi_{V_2}(2\omega_s)\Big],$$

$$m_2 = (\frac{\omega_s^2 - \omega_2^2}{2\omega_s})\tilde{X}_1 - \zeta_2\omega_2\tilde{X}_2 + \frac{\gamma\omega_s}{8}\Big[(\tilde{X}_1^2 + 3\tilde{X}_2^2)\ x$$

$$\cos\ \Psi + 2\tilde{X}_1\tilde{X}_2\ \sin\ \Psi\Big] + \frac{\gamma}{\omega_s}\ Z_1\ \cos\ \Psi + \frac{\pi\gamma^2}{2}\ \tilde{X}_2\Big[\Phi_{V_2}(0)$$

$$+ \Phi_{V_2}(2\omega_s)\Big],$$

$$(\underline{\sigma}\ \underline{\sigma}')_{11} = \frac{\pi\gamma^2}{4}\{\Big[2\Phi_{V_2}(0) + \Phi_{V_2}(2\omega_s)\Big]\tilde{X}_1^2 + \tilde{X}_2^2\ \Phi_{V_2}(2\omega_s)\}$$

$$+ \frac{\pi\gamma^2}{\omega_s^2}\ \Phi_\xi(\omega_s),$$

$$(\underline{\sigma}\,\underline{\sigma}')_{22} = \frac{\pi\gamma^2}{4}\left\{\tilde{X}_1^2\,\phi_{V_2}(2\omega_s) + \tilde{X}_2^2\left[2\phi_{V_2}(0) + \phi_{V_2}(2\omega_s)\right]\right\} + \frac{\pi\gamma^2}{\omega_s^2}\,\phi_\xi(\omega_s),$$

$$(\underline{\sigma}\,\underline{\sigma}')_{12} = (\underline{\sigma}\,\underline{\sigma}')_{21} = \frac{\pi}{2}\,\gamma^2\,\tilde{X}_1\tilde{X}_2\,\phi_{V_2}(0). \tag{4.55}$$

The Itô equation for the system is

$$d\begin{Bmatrix}\tilde{X}_1\\ \tilde{X}_2\end{Bmatrix} = \begin{Bmatrix}m_1\\ m_2\end{Bmatrix} dt + \begin{bmatrix}\sigma_{11} & \sigma_{12}\\ \sigma_{21} & \sigma_{22}\end{bmatrix}\begin{Bmatrix}dB_1\\ dB_2\end{Bmatrix} \tag{4.56}$$

From (4.55) one can also construct the Fokker-Planck equation for the transition probability density for $\tilde{X}_1$ and $\tilde{X}_2$, but its solution appears very difficult.

To investigate the second moments of the response, one applies Itô's differential rule to X_1^2, X_1X_2 and X_2^2 and takes the expectations:

$$\frac{d}{dt}\begin{Bmatrix}E[\tilde{X}_1^2]\\ E[\tilde{X}_1\tilde{X}_2]\\ E[\tilde{X}_2^2]\end{Bmatrix} = \begin{bmatrix}a+5b & -2c & b\\ c & a+4b & -c\\ b & 2c & a+5b\end{bmatrix}\begin{Bmatrix}E[\tilde{X}_1^2]\\ E[\tilde{X}_1\tilde{X}_2]\\ E[\tilde{X}_2^2]\end{Bmatrix} + \begin{Bmatrix}1\\ 0\\ 1\end{Bmatrix}\frac{\pi\gamma^2}{\omega_s^2}\,\phi_\xi(\omega_s) + \begin{Bmatrix}r_1\\ r_2\\ r_3\end{Bmatrix}, \tag{4.57}$$

where $a = -2\zeta_2\omega_2 + \frac{3}{2}\pi\gamma^2\phi_{V_2}(0)$,

$\quad b = \frac{1}{4}\pi\gamma^2\phi_{V_2}(2\omega_s)$,

$\quad c = (\omega_s^2 - \omega_2^2)/2\omega_s$,

$$r_1 = \frac{\gamma\omega_s}{4}\{E[3\tilde{X}_1^3 + \tilde{X}_1\tilde{X}_2^2]\sin\psi + 2E[\tilde{X}_1^2\tilde{X}_2]\cos\psi\}$$

$$+ \frac{2\gamma}{\omega_s}E[Z_1\tilde{X}_1]\sin\psi,$$

$$r_2 = \frac{\gamma\omega_s}{8}\{E[5\tilde{X}_1^2\tilde{X}_2 + \tilde{X}_2^3]\sin\psi + E[5\tilde{X}_1\tilde{X}_2^2 + \tilde{X}_1^3]\cos\psi\}$$

$$+ \frac{\gamma}{\omega_s}\{E[Z_1\tilde{X}_2]\sin\psi + E[Z_1\tilde{X}_1]\cos\psi\},$$

$$r_3 = \frac{\gamma\omega_s}{4}\{E[\tilde{X}_1^2\tilde{X}_2 + 3\tilde{X}_2^3]\cos\psi + 2E[\tilde{X}_1\tilde{X}_2^2]\sin\psi\}$$

$$+ \frac{2\gamma}{\omega_s}E[Z_1\tilde{X}_2]\cos\psi. \tag{4.58}$$

One finds that these equations involve the third order moments as well as $E[Z_1\tilde{X}_1]$ and $E[Z_1\tilde{X}_2]$.

To obtain some useful solutions for the present problem we shall restrict ourselves to the resonance case where $\omega_s = \omega_s$. Furthermore, advantage is taken of a clue from a companion deterministic analysis[28], which suggests that equation (4.56) may be linearized about a "sink" ($X_o \sin \psi$, $X_o \cos \psi$) of the solution vector field. Specifically, we let

$$\tilde{X}_1 = X_o \sin \psi + S_1$$
$$\tilde{X}_2 = X_o \cos \psi + S_2 \qquad (4.59)$$

and replace functions of $\tilde{X}_1$ and $\tilde{X}_2$ by

$$f_1(\tilde{X}_1) = f_1(X_o \sin \psi) + S_1 f_1'(X_o \sin \psi)$$

$$f_2(\tilde{X}_2) = f_2(X_o \cos \psi) + S_2 f_2'(X_o \cos \psi),$$

$$f_3(\tilde{X}_1, \tilde{X}_2) = f_3(X_o \sin \psi,\ X_o \cos \psi) +$$

$$S_1 \left[\frac{\partial}{\partial \tilde{X}_1} f_3\right]_{\tilde{X}_1 = X_o \sin \psi,\ \tilde{X}_2 = X_o \cos \psi} +$$

$$S_2 \left[\frac{\partial}{\partial \tilde{X}_2} f_3\right]_{\tilde{X}_1 = X_o \sin \psi,\ \tilde{X}_2 = X_o \cos \psi}$$

Substituting these into equation (4.56) and equating the terms not involving S_1 nor S_2 in the expressions for m_1 and m_2 to zero, we obtaine

$$X_o = \frac{2}{3\gamma\omega_2}\left[\bar{\alpha} - (\bar{\alpha}^2 - 6\gamma^2 Z_1)^{1/2}\right] \qquad (4.60)^*$$

* The other possible solution for X_o is a saddle point; hence it is discarded for being unsuitable as a centre for linearization.

where

$$\bar{\alpha} = 2\zeta_2\omega_2 - \pi\gamma^2\left[\Phi_{V_2}(0) + \Phi_{V_2}(2\omega_2)\right]$$

and the linearised version of equation (4.56) as follows:

$$d\begin{Bmatrix} S_1 \\ S_2 \end{Bmatrix} = \begin{Bmatrix} \bar{m}_1 \\ \bar{m}_2 \end{Bmatrix} dt + \begin{bmatrix} \bar{\sigma}_{11} & \bar{\sigma}_{12} \\ \bar{\sigma}_{21} & \bar{\sigma}_{22} \end{bmatrix} \begin{Bmatrix} dB_1 \\ dB_2 \end{Bmatrix} \tag{4.61}$$

In equation (4.61),

$$\bar{m}_1 = \left\{-\zeta_2\omega_2 + \frac{\gamma\omega_2}{4}(2 - \cos 2\psi)X_o + \frac{\pi\gamma^2}{2}\left[\Phi_{V_2}(0) + \Phi_{V_2}(2\omega_2)\right]\right\}S_1 + \frac{\gamma\omega_2}{4}(\sin 2\psi)X_o S_2$$

$$\bar{m}_2 = \frac{\gamma\omega_2}{4}(\sin 2\psi)X_o S_1 + \left\{-\zeta_2\omega_2 + \frac{\gamma\omega_2}{4}(2 + \cos 2\psi)X_o + \frac{\pi\gamma^2}{2}\left[\Phi_{V_2}(0) + \Phi_{V_2}(2\omega_2)\right]\right\}S_2$$

$$\left[\underline{\bar{\sigma}}\ \underline{\bar{\sigma}}'\right] = \left[\underline{\sigma}\ \underline{\sigma}'(X_o\sin\psi,\ X_o\cos\psi)\right]$$

$$+ S_1\left(\frac{\partial}{\partial\tilde{X}_1}\underline{\sigma}\ \underline{\sigma}'\right)_{\tilde{X}_1 = X_o\sin\psi} + S_2\left(\frac{\partial}{\partial\tilde{X}_2}\underline{\sigma}\ \underline{\sigma}'\right)_{\tilde{X}_2 = X_o\cos\psi}. \tag{4.62}$$

The condition for the first moment stability is found to be

$$2\zeta_2\omega_2 > \gamma\sqrt{6}\, Z_1 + \pi\gamma^2\left[\Phi_{V_2}(0) + \Phi_{V_2}(2\omega_2)\right]. \tag{4.63}$$

It is of interest to note that in the absence of turbulence, $\phi_{V_2}(\omega) = 0$, inequality (4.63) reduces to the same result obtained in a deterministic analysis, reference 28.

APPENDIX I

STRATONOVICH-KHASMINSKII LIMIT THEOREM

It has been known for some time that in certain random vibration calculations the excitation process may be idealized as a white noise even though white noise processes are physically unrealizable. For example, in the case of a single-degree-of-freedom linear oscillator under the excitation of a broad-band noise, the mean-square deflection can be computed from

$$E\left[X^2\right] = \frac{\pi}{2m\zeta\omega_o^3} \phi_F(\omega_o)$$

where m, ζ, ω_o are the usual symbols for mass, damping ratio, and the natural frequency, respectively, and $\phi_F(\omega_o)$ is the spectral density of the excitation evaluated at the system natural frequency ω_o. The implications are that the excitation process is replaceable by a white noise with a constant spectral density equal to $\phi_F(\omega_o)$ and that the response of the system, when represented by a vector $(X,\dot{X})$ in the phase-plane, is close to a Markov vector in some sense. The substitution is valid provided that the original spectrum of the excitation is broad and varies slowly in the neighbourhood of the system natural frequency ω_o.

In a more complicated situation, especially one involving parametric excitations or nonlinear relationships, the substitution of actual excitations by white noise processes is not a simple task and it can lead to erroneous results if not done properly. The Stratonovich-Khasminskii limit theorem provides a mathematical framework to carry out such substitutions and to allow for necessary corrections which must be taken into account when actual excitations are replaced by white noise processes (or differentials of Wiener processes).

The theorem applies to the case governed by the differential equation

$$\dot{X} = \varepsilon\, f(X,t) + \varepsilon^{1/2} g(X,t)\eta(t), \qquad |\varepsilon| << 1$$

where $\eta(t)$ is a stochastic process with a zero mean and a correlation time

$$\tau_{cor} << O(\frac{1}{\varepsilon})$$

Then X(t) converges weakly to a Markov process $X^*(t)$ governed by the Itô equation

$$dX^* = \varepsilon m(X^*)dt + \varepsilon^{1/2}\sigma(X^*)dB$$

where B(t) is a unit Wiener process, and m and σ are

computed from

$$m = \lim_{T\to\infty} \frac{1}{T}\left[\int_0^T f(X,t)dt + \int_0^T dt \int_{-\infty}^{0} g(X,t+\tau)\right.$$

$$\left.\frac{\partial g(X,t)}{\partial X} R_\eta(\tau)d\tau\right]$$

$$\sigma^2 = \lim_{T\to\infty} \frac{1}{T}\left[\int_0^T dt \int_{-\infty}^{\infty} g(X,t)g(X,t+\tau)R_\eta(\tau)d\tau\right]$$

In the case where X is a vector, m is generalized to a column matrix and σ^2 is generalized to the product $\underline{\sigma}\underline{\sigma}'$ where $\underline{\sigma}$ is a square matrix.

APPENDIX II

THE ITÔ DIFFERENTIAL RULE

Let X(t) be a Markov process governed by the Itô equation

$$dX = m(X,t)dt + \sigma(x,t)dB$$

and let $\phi(X,t)$ be an arbitrary function of X which satisfies certain differentiability conditions. Then

$$d\phi = (\frac{\partial\phi}{\partial t} + \mathcal{L}_X\phi)dt + \sigma(X,t)\,\frac{\partial\phi}{\partial X}\,dB$$

where $\mathcal{L}_X$ is the backward operator defined as

$$\mathcal{L}_X = m(X,t)\,\frac{\partial}{\partial X} + \frac{1}{2}\,\sigma^2(X,t)\,\frac{\partial^2}{\partial x^2}$$

The generalization of the Itô differential rule to a Markov vector is immediate.

REFERENCES

[1] Sears, W.R., Some aspects of nonstationary airfoil theory and its practical application, Journal of Aeronautical Sciences, 8, 104, 1941.

[2] Liepmann, H.W., On the application of statistical concepts to the buffeting problem, Journal of Aeronautical Sciences, 19, 793, 1952.

[3] Howell, L.J. and Lin,Y.K., Response of flight vehicles to nonstationary, atmospheric turbulence, AIAA Journal, 9, 2201, 1971.

[4] Fung, Y.C., Statistical aspects of dynamic loads, Journal of Aeronautical Sciences, 20, 317, 1953.

[5] Lin, Y.K., Transfer matrix representation of flexible airplanes in gust response study, Journal of Aircraft, 2, 116, 1965.

[6] Fujimori, Y. and Lin,Y.K., Analysis of airplane response to nonstationary turbulence including wing

bending flexibility, AIAA Journal, 11, 334, 1973.

[7] Corcos, G.M., Resolution of pressure in turbulence, Journal of Acoustical Society of America, 35, 192, 1963.

[8] Lin, Y.K. and Maekawa, S., Decomposition of turbulence forcing field and structural response, AIAA Journal, 15, 1977.

[9] Morse, P.M. and Ingard, K.U., Theoretical Acoustics, McGraw-Hill Book Company, New York, 707, 1968.

[10] Lin, Y.K., Maekawa, S., Nijim, H. and Maestrello, L., Response of periodic beam to supersonic boundary-layer pressure fluctuations, presented at IUTAM Symposium on Stochastic Problems in Dynamics, University of Southampton, England, 1976.

[11] Van der Hoven, I., Power spectrum of horizontal wind speed in the frequency range from 0.0007 to 900 cycles per hour, J.Meteorology, 14, 160, 1957

[12] Archibald, E.D., Some results of observations with kite-wire suspended anemometers up to 1300 feet above

the ground in 1883-1885, Nature, 33, 593, 1886.

[13] Davenport, A.G., A rationale for determination of design wind velocities, Proceedings, A.S.C.E., J. Structural Div., 86, 39, 1960.

[14] Davenport, A.G., The treatment of wind loading on tall buildings , Proceedings, Symposium on Tall Buildings, University of Southampton, Pergamon Press, 1966.

[15] Gumbel, E.J., Statistical of Extremes, Columbia University Press, New York, 1958.

[16] Davenport, A.G., The application of statistical concepts to wind loading of structures, Proceedings, Institution of Civil Engineers, 19, 449, 1961.

[17] Berman, S., Estimating the longitudinal wind spectrum near the ground, Qu. J. Roy. Mer. Soc. 91, 302, 1965.

[18] Vaicaitis, R., Shinozuka, M. and Takeno, M., Parametric study of wind loading on structures, Proceedings, A.S.C.E., J.Structural Eng., ST3, 453, 1973.

[19] Etkin, B., Theory of the response of a slender vertical structure to a turbulent wind with shear, presented at the Meeting on Ground Wind Load Problems in Relation to Launch Vehicles, NASA Langley Research Center, U.S.A., 1966.

[20] Stratonovich, R.L., Topics in the Theory of Random Noise, Vol.II, translated by R.A.Silverman, Gordon and Breach, 1967.

[21] Khasminskii, R.Z., A limit theorem for the solution of differential equations with random right hand sides, Theory of Probability and Applications, 11, 390, 1966.

[22] Bogoliubov, N.N. and Mitropolskii, Y.A. Asymptotic Methods in the Theory of Nonlinear Vibration, Gordon and Breach, 1961.

[23] Itô, K., On stochastic differential equations, Memoirs of the American Mathematical Society, 4, 51, 1951.

[24] Ariaratnam, S.T., Stability of mechanical systems under stochastic parametric excitations, in Lecture Notes in Mathematics No.294, ed. R.F. Curtain, Springer-

Verlag 291, 1972.

[25] Infante, E.F., On the stability of some linear non-autonomous systems, Journal of Applied Mechanics, 35, 7, 1968.

[26] Kozin, F. and Wu, C.M., On the stability of linear stochastic differential equations, Journal of Applied Mechanics, 40, 87, 1973.

[27] Ariaratnam, S.T. and Tam, D.S.F., Parametric Excitation of a damped Mathieu oscillator, ZAMM, 56, 449, 1976.

[28] Holmes, P. and Lin, Y.K., to be published.

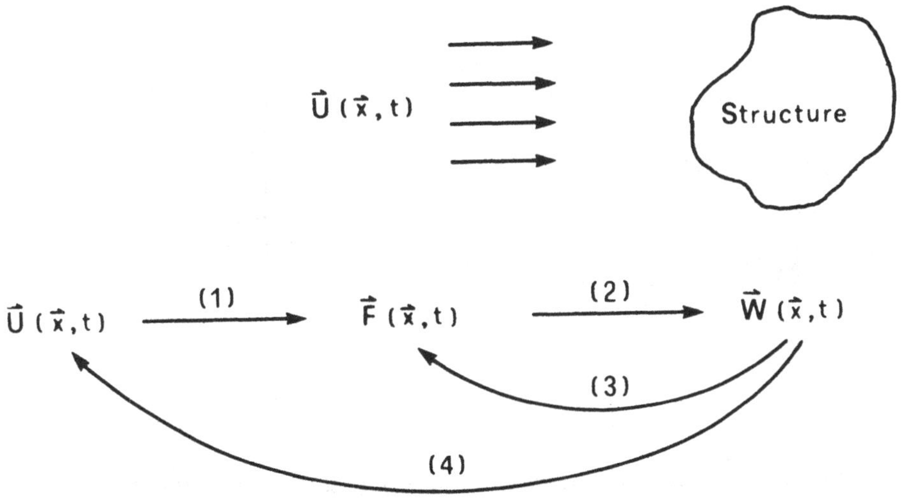

Fig. 1 - 1 Fluid-Structure Interaction

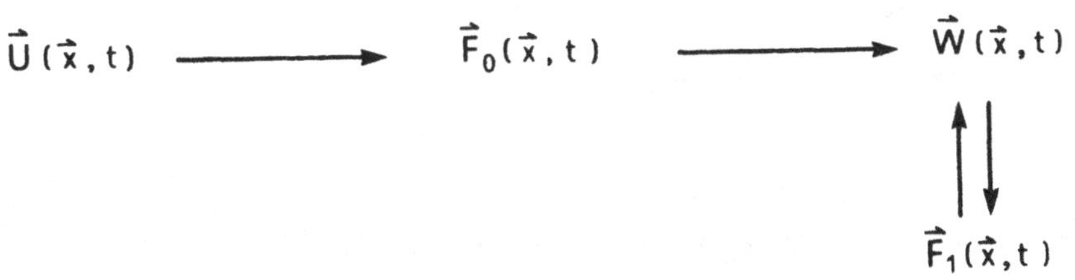

Fig. 1 - 2 A Linearization Scheme

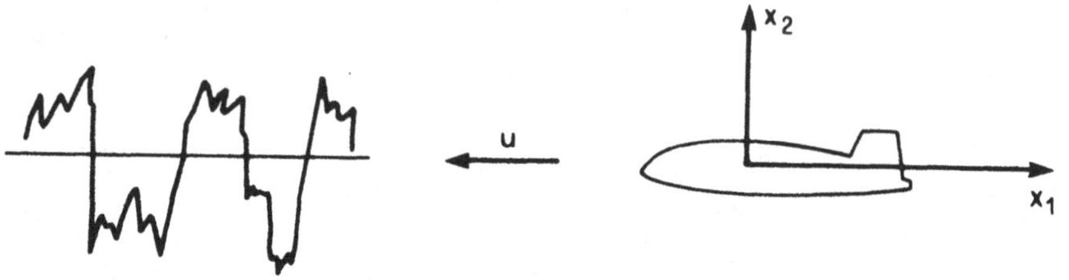

Fig. 2 - 1 An Airplane Entering a Frozen Pattern of Random Gust Field

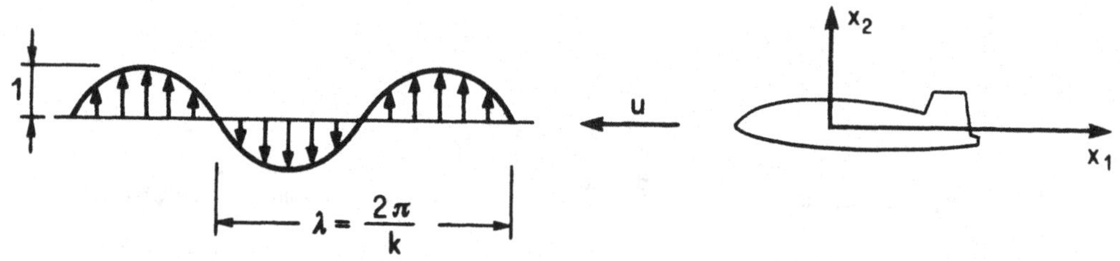

Fig. 2 - 2 An Airplane Entering a Frozen Pattern of Sinusoidal Gust Field

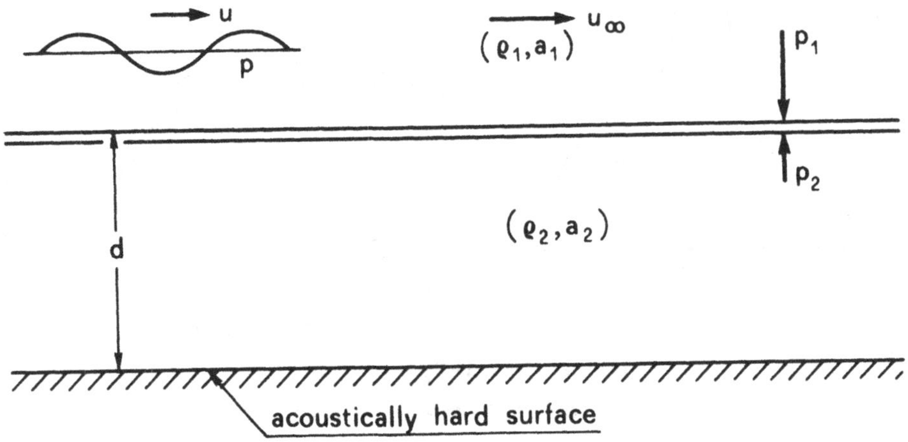

Fig. 3 - 1 An Infinite Beam under the Excitation of a Convected Sinusoidal Pressure Field

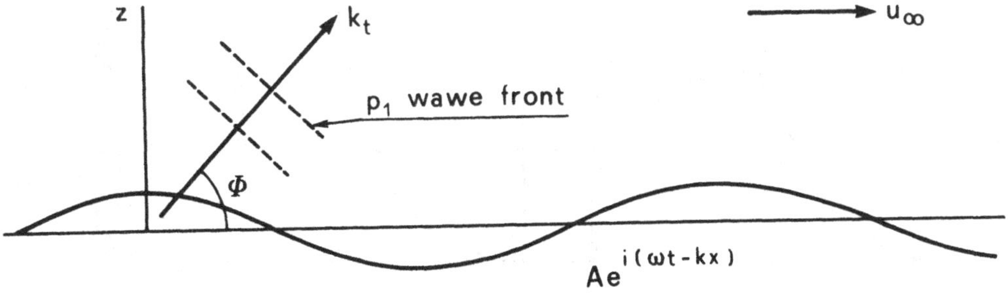

Fig. 3 - 2 Acoustic Pressure p_1 Induced by Structural Motion

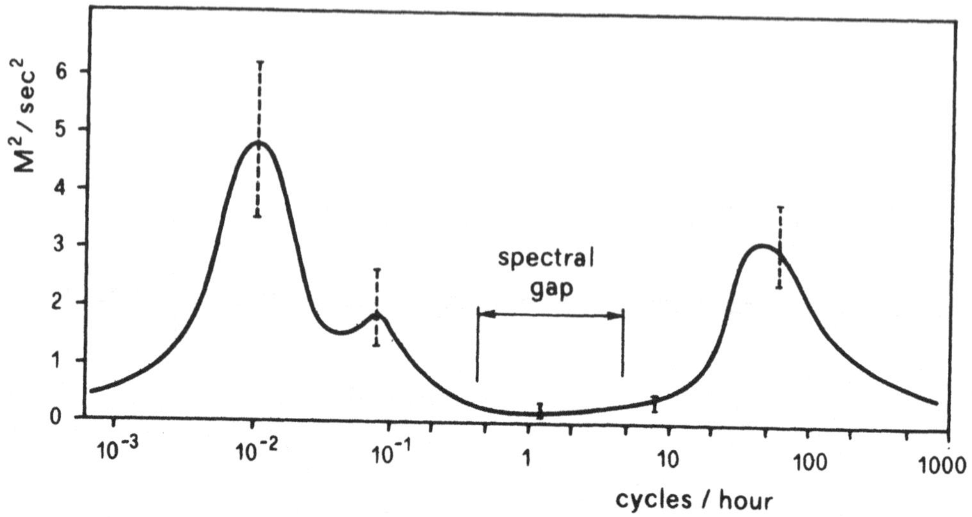

Fig. 4 - 1 Horizontal Wind-Speed Spectrum at Brookhaven National Laboratory at about 100-m height (reference 11)

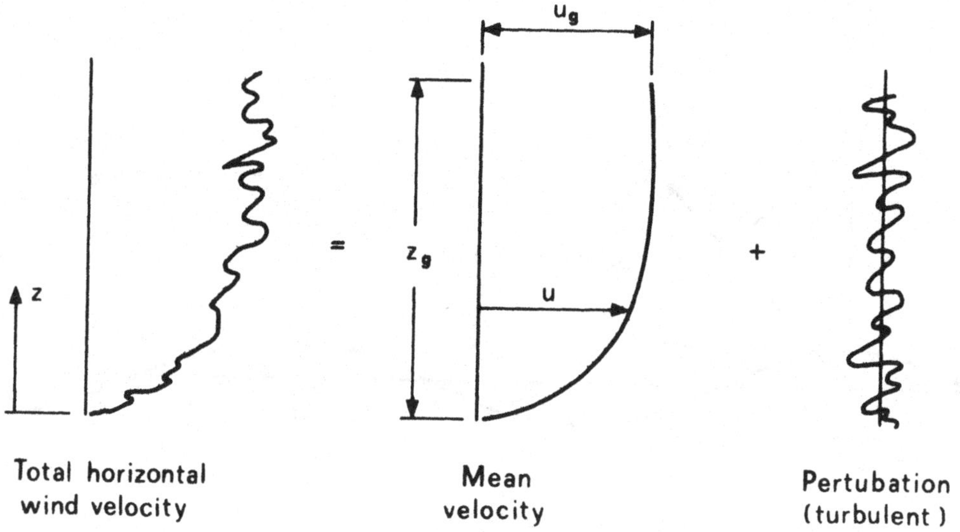

Fig. 4 - 2 Wind Velocity near the Ground

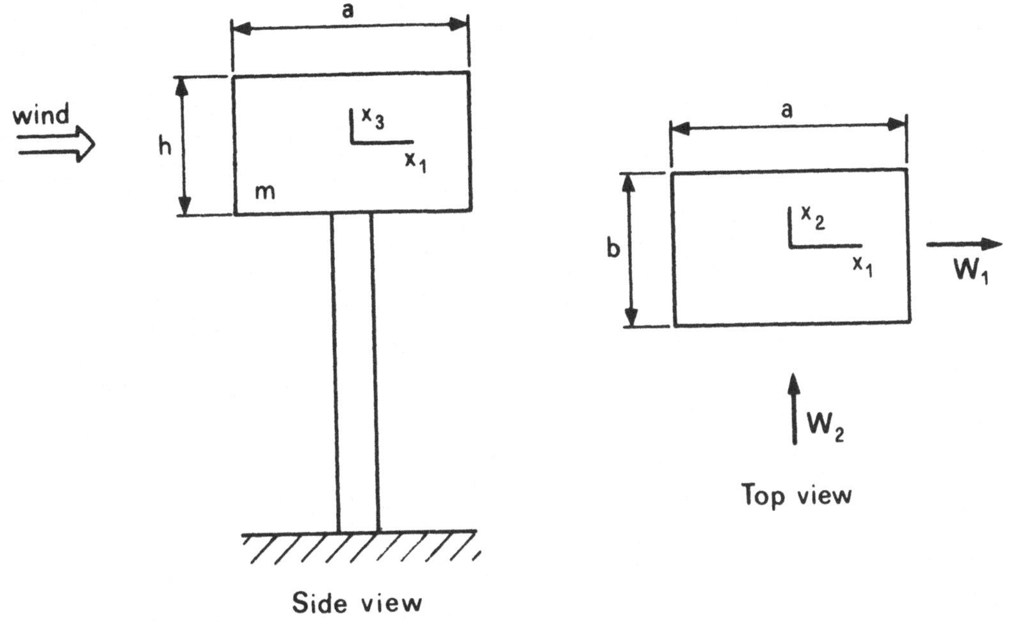

Fig. 4 - 3 Simplified Model of a Ground Structure under Wind Excitation

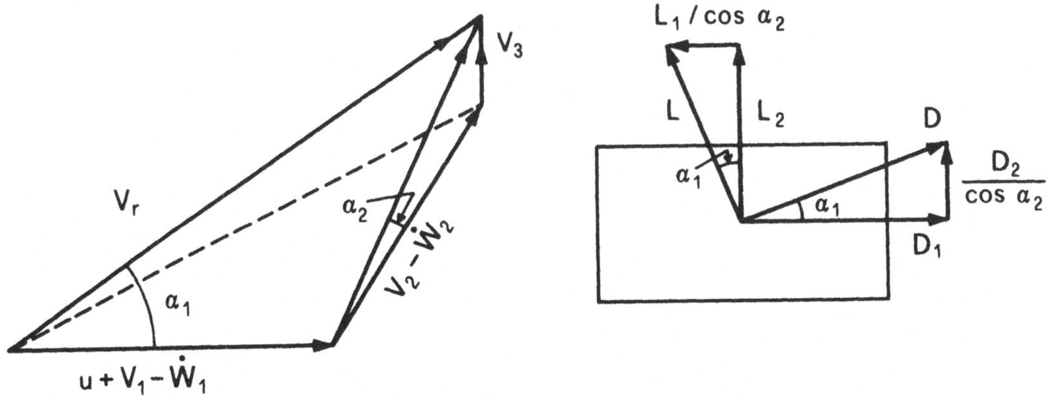

Fig. 4 - 4 Components of Wind Velocities Relative to the Structure and Induced Forces

Printed in the United States
By Bookmasters